The Marketer's Concise Guide to

CRO *

Tips, Tests, and Tactics to Gather More Leads and Grow Your Business

*Conversion Rate Optimization

by Scott Frangos

ISBN-13: 978-0-9969504-0-4 | **ISBN-10**: 0996950400
(Paperback Edition. eBook versions also available)
Published by: Webdirexion Publishing, a division of *Webdirexion LLC* 13215
SE Mill Plain Blvd. • Suite C8 #264 • Vancouver, WA 98684 • **888.974.9522**

Executive Author/Cover Design: Scott Frangos
Managing Editor: Margot Hall
Copy Editor: Whitney Beyer
Contributors: Julie Hume, Miranda Booher & Sherri Gutierrez

Additional Copies of this Book: This book is available online for print and ebook orders. Instructors, colleges, & Independent bookstores may arrange for wholesale ordering via: Webdirexion.com/CRO-Guide

CRO Guide Listings: Contact publisher about paid listings in the back of the in future editions.

Disclosure: *Webdirexion* has marketing agreements with some, but not all, of the companies mentioned in this book, and links in the books section are via affiliate codes that provide a small monetary reward. Our goal is to provide you with a concise but thorough reference to leading services and products for marketing professionals. We have read each and every book we recommend in this Guide.

What Others are Saying about the CRO Guide:

You wanna know something interesting? I've worked with Scott. I know the value of testing. Yet I get lazy. I justify not testing because "the site doesn't have enough traffic" or I become complacent thinking my site is good enough. This quick read reminded me how important it is to continually test – the market moves too quickly to do anything else. And, as I read, I took notes for a half dozen tests my clients deserve. Thanks, Scott, for the motivation and tools to be better at my job. — Joe Hage, CEO, *Medical Marcom*

Scott and his team did an excellent job of researching and communicating every facet of conversion rate optimization. This is a topic that I would not consider a "page turner," but they've made it interesting. My biggest takeaway is the need to test every pitch, every call to action and every layout. Not even the most talented, experienced creatives know what will work and what won't. Even Don Draper didn't know (dirty little secret, they tested back in the '60s too). This book will make you rich. — Bob Leonard, Managing Consultant, *acSellerant*

Scott and the Webdirexion *team created the most comprehensive guide on conversion rate optimization I have read. It covers everything you need to know to master CRO. It contains hands-on, practical tips. I wanted to stop reading (but couldn't) and get busy reviewing all my landing pages. I'd recommend the CRO guide to anyone who wants to learn how to boost 'connections' with visitors. This guide will help you create more engaged readers and turn them into sales. In the end you will become a smarter marketer.* — Bill Flitter, CEO, *Dlvr.it*

Scott Frangos has done a masterful job explaining the finer points of Conversion Rate Optimization (CRO). For our business, we've never had a problem with the content piece, but rather the testing and the closing part. After reading Scott's book, I now feel way more confident that we can improve our CRO efforts. The book is a solid read and a quick one as well. Highly recommended! — Jon Wuebben, CEO *Content Launch*, Author, *Content is Currency*

*For my parents and family who have always stood
behind me, for my friends who stood the test,
for my colleagues and mentors upon whom
I have stood, for the Webdirexion team who continue
to inspire and stand for measurable results, for our
companion animals who stand by us no matter what, and
to all our readers who wish to succeed in marketing.*

Table of Contents

Foreword

In 2003, I read an article in a B2B magazine from Don Schultz, the father of integrated marketing. In it, Don states that everything you do as a company can now be duplicated (rather easily) by another company. The advances in technology have enabled this to happen, and that rather quickly.

If this is true, how do we differentiate our brand? Don says that the key is how we communicate with our prospects and customers. So simple, yet so true.

That's why the content you create and distribute is so critical for your business (not just your marketing). If we can build an audience that knows, likes, and trusts us, we tip the scales in our favor for that person to become a customer...or better yet, turn so-so customers into amazing advocates for the business.

But to this point, the results aren't great. In the Content Marketing Institute/MarketingProfs 2016 Content Marketing Benchmark study, content marketing effectiveness actually went down year over year. Whether you are a B2B, B2C, nonprofit, small business, or large enterprise, effectiveness rates are less than 40 percent.

One of the main reasons why is that organizations are publishing all over the place, without any real strategy or execution plan that works for business outcomes.

And this is exactly why the book you are reading right now is so important. Yes, target a specific audience with your content. Identify a content niche where you can actually tell a different story than everyone else. Convert subscribers into leads, then prospects, and then customers. But do it in an intelligent, efficient way.

I see so many businesses set off to build a loyal audience through content, but just don't do the little things that convert more passersby to subscribers and, ultimately, people interested in your products and services. Why go through all the trouble and let them slip away like "sand through the hourglass"?

You have all the tools and data to make this happen. You just need to commit to making the decision to *actually communicate more effectively*. Don't your customers deserve that? Doesn't your business deserve that? Now is your time. Make the decision to take advantage of this amazing opportunity.

— *Joe Pulizzi*

Founder, *Content Marketing Institute*; Author, *Content Inc: How Entrepreneurs Use Content to Build Massive Audiences and Build Remarkably Successful Businesses*

Introduction

This is the guide that I wish I'd had when I started building websites and applying marketing and advertising principles earlier in my career. Most clients still ask how to get more visitors to their websites. That question begins with the notion of "hits" — as if only quick looks mattered. A better question is: "how do we get more qualified visitors to our websites?" And better still, "how do we get qualified website visitors to take the actions [conversions] we want on our sites?" Ask that question and you're begging for a CRO answer.

Is Content King? It's more like the powerful Queen, zipping everywhere on the board, interacting with all the pieces. But the game is not over until you mate (Convert) with the King—the final Connection.

The Three Cs of Marketing: Many of today's marketers are focused primarily on content, or "content marketing." While I agree that content goes to the heart of your online publishing effort and that it can engage the right prospects, I don't think it's king. Maybe queen, like the all-powerful piece in the game of chess. The chess queen is very agile, moving all around the board, just like content popping up in ads, social media, white papers, and even in old technology for printed marketing collaterals.

Why is content not king? Simply put, because great content alone does not close sales. You've got to get past the queen, and shake hands with the final deal maker.

You'll find some useful formulas in the 25 CRO Tips section at the end of this book, but let me offer one here that illustrates my point:

Content + Conversion = Connection

And it is only after a connection that you make a sale.

CRO involves an interesting commingling of left-brain and right-brain strategies and tactics that go beyond just good content creation.

That's why CRO tactics are so needed in our industry. If you stop only with content, you might think spending your day robo-posting links to blog posts via social media is all you have to do. There's more.

If I have a trade show booth, I need more than just signage and a pile of brochures. I need to do more than just get on the PA system or Twitter and shout out my booth number. I need real people to represent our company, people on the ground who are persuasive enough to create an emotional connection with prospects in order to convert them to leads.

Of course, there's always content in marketing before and after conversions, in products and services, and in all the stories we tell on our journeys. And, what is content? In the end, it is the truth that comes from the heart and leads to a real and valuable connection. I am so grateful for the connections I have had with content and conversion mentors and thought leaders such as Joe Pulizzi, Russell Sparkman, Bill Flitter, Joe Hage, Bob Leonard, New Barret, and authors and teachers including Ardith Albee, Tim Ash, Flint McGloughlin, Rand Fishkin, Robert Cialdini, Michael Procopio, and Brian Massey.

Then there's our *Webdirexion* team of talented and passionate marketers and fascinating world travelers who helped produce this guide. Their heartfelt experience and learning has made this guide richer. They've taught me too. Margot, Whitney, Julie, Miranda, Sherri—you're all great, and you have my thanks.

We've developed this guide to give you a concise overview of technique and tool options with enough examples from the trenches to get you going. In the back of the book are a couple of brainstorming forms to help you lock in your learning and brainstorm your next tests. We also have an appendix with a review of four cloud-based testing services to help you choose which one is best for you. Our goal is to give you a shorter read so you can quickly start implementing the ideas herein to help you grow leads.

Now, let's talk conversions. The game is not over until you capture the king.

— Scott Frangos, October 10, 2015

1

Why is Conversion Rate Optimization Important?

3 "C's": Content + Conversion = Connection, and it is only after a connection that you make a sale.

After that you have the golden "C" — a new Customer.

Chapter 1. Why is CRO Important?

Two questions: How much is a new customer worth to you? And, how valuable is a website and associated marketing tactics that continually gather more leads to convert visitors into customers? The answer to the first question is the value of the first sale + the revenue from continuing business with that customer + referrals you might get from an extremely satisfied customer. In other words, new customers are golden.

When determining the value of a website and associated marketing tactics used to gather leads and convert visitors into customers, most marketers fall flat. Sure, most businesses and smart marketers endeavor to build websites to attract visitors, generate leads, and increase their business' bottom line, with the ultimate goal of converting visitors into customers. But how can they be more effective during this process and win more customers? Enter conversion rate optimization, or CRO.

CRO adds testing methods to iteratively improve your results, taking the guesswork out of the process. Forty years ago there was an article in Reader's Digest that described how to build a mechanical computer to play a watered-down, simple game of chess using only three pawns. You programmed it to win, by first placing all possible choices into the computer, then removing all the bad choices. This is what you do in CRO — test to remove the bad choices of elements on a page or in your site that are not working or being "chosen" (usually the choice is indicated with a click) by your visitors.

Too often, a web designer may not be thinking like a customer going through your site. A designer might set the page up for a visitor to flow logically and sequentially through the site or landing page, when in reality the visitor needs an intuitive, non-sequential buying sequence. Only iterative testing via CRO helps you eliminate the bad moves and provide the good moves to persuade a visitor to convert to a lead.

Here are two equations:

First Design = Best Guess

Sure you can, and should, test your first design even before coding (we'll tell you how), but until real visitors encounter your actual site and landing pages, you are still guessing at what they will do and why.

CRO = ROI + Smiles

When you gather more *qualified* leads, you get smiles from your sales team. When your sales team connect with your leads and convert them into new customers, you get both a return on your testing investment and smiles from the C-Level people in your company. What could be better?

Now, let's take a look at the CRO process itself.

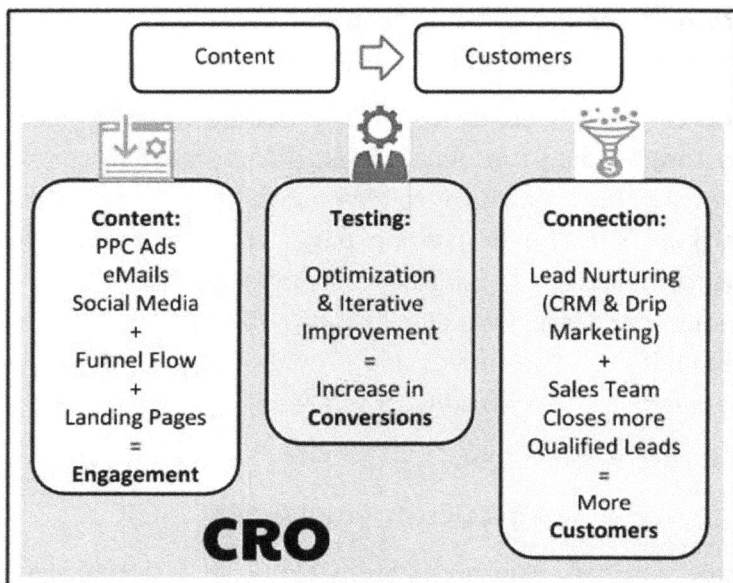

| Content | ⇨ | Customers |

Content:
PPC Ads
eMails
Social Media
+
Funnel Flow
+
Landing Pages
=
Engagement

Testing:
Optimization
& Iterative
Improvement
=
Increase in
Conversions

Connection:
Lead Nurturing
(CRM & Drip
Marketing)
+
Sales Team
Closes more
Qualified Leads
=
More
Customers

CRO

Testing provides an often missing link between content and connection. A lot of marketers focus on the top row (Content to Customers), leaving out testing tactics, because their training focuses on content creation. When you become trained in CRO, you will increase your leads and gain more customers.

CRO is the process of improving your site's performance with the help of testing tactics, analytical review, and the iterative improvements you make to increase connection. In other words, it's the skillful art and measured science of refining your website to speak directly to the audience you are trying to reach. CRO allows you to use existing site traffic to get more of the results you want, which will ultimately increase return on investment (ROI). But before we jump into the whats and hows, maybe we ought to start with the whys.

First, the obvious: Why not practice CRO? We have the tools and technology to measure site performance, and such metrics extend to every key point in a website visitor's experience. It's true that people have found success through a preemptively measured approach, by anticipating consumer behavior based on general consumer and demographic information. But why follow a hunch when you can give your target market exactly what they want? We think that since so many on marketing-communications teams are trained at content creation, they don't yet feel comfortable with testing for conversion rate improvements.

On average, for every $92 a business invests into bringing visitors to its site, they invest only $1 into converting them. Sure, getting people to visit your site is important, but it's crucial to keep them there long enough to convert them into leads. The typical online consumer's attention span is fleeting, and the many distractions of the Internet—banner ads, hyperlinks, pop-ups, etc.—only serve to further fracture our capacity for focus.

The bottom line is this: If you're going to spend money to garner web traffic, you should take the necessary steps to ensure that you're attracting and keeping the attention of the right visitors. Marketers are often so concerned with creating great content that they fail to focus on what happens after visitors engage with their content. Improving conversion rates leads to authentic connections, which, in turn, grow sales. Simply put, real content matters, but it can only take you so far—it's connections that convert. It takes some extra time, but it almost always pays off. CRO is a great investment.

At its core, CRO is about testing, retesting, and—you guessed it—testing again. It's about making small yet significant improvements to multiple aspects of your website, thereby refining your visitors' experiences into an efficient converting machine. Testing can be tedious business, but taking the time to get it right will help you maximize your initial investment.

If you are not writing at least three versions of the ads at the beginning of each pay-per-click campaign to find the best converting ad, then you're leaving money at the table. If you fail to test your landing pages, pop-up calls to action (CTAs), sign-up forms, email subject lines, and sales follow-up letters, who can say just how much you are losing?

The idea of testing for and tracking all that data can be intimidating, which is likely one of the reasons that so many marketers overlook the potential that testing has to boost the performance of their sites. Don't let that hold you back. Taking a measured approach will allow you to get the most out of the resources you pour into your site and, ultimately, boost your bottom line. After all, isn't that the point?

2

Types of Testing

Testing goes beyond analytics, which tells what visitors do and what they don't do, to help you find out what really does and doesn't work on landing pages. Testing enables you to eventually gather more leads and to improve how people flow through your site toward desired business goals.

Chapter 2. Types of Testing

When it comes to testing, you need to start with a firm grasp of your business objective. Is it to pull downloads of a white paper? Are you looking for more video views? Do you need to capture leads?

Once your goals are defined, ask yourself one more question: Is reviewing analytics enough for smart marketing corrections? The short answer is no. That's because analytics tell you what people do (i.e., how many pages they visit) or do not do (i.e., when they bounce), but analytics do not and cannot tell you *why*. CRO tests can help you understand the intentions of your prospective leads much better in order to determine the *why* of their behavior.

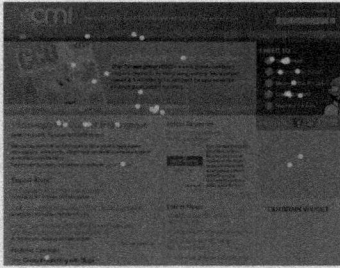

Click Test (before coding a website)

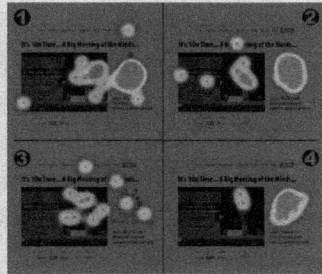

Heat Map Test

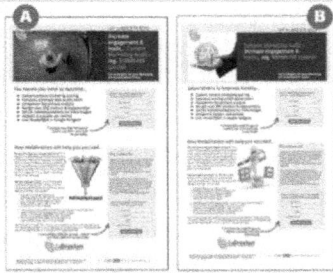

A/B Test

Nav Flow Test

(Left to right): In the click test, we asked reviewers to click on elements that would best serve content marketers before we coded the design to help us make some last minute revisions for the Content Marketing Institute. *The heat map test shows which of four CTAs might be most compelling (see* Chapter 4*). The A/B test shows two versions of a page where we varied headlines, images, and CTAs to find the best lead-gathering results. The nav flow test shows how well people move through a series of web pages to get to and complete a desired goal.*

Testing before you code

Most web projects will initially involve showing a wireframe, and then a design composite image—usually a JPEG—of a proposed site. But very few marketers take the short time required to run the design past a number of objective reviewers. Even just asking, "What is this site about?" after they look at the to-size JPEG design image will give your design team key feedback *before* coding. Remember: your designer, marketing director, and any C-level execs are only guessing at what will work without approaching a new design or website update with the eyes that really matter—independent prospects. For more details, see *Chapter 4: Testing Before You Code a Page or Site*.

Heat map and click tracking tests

Heat maps may be used before (usually on a JPEG image of a site page or interface) or after you code your pages and sites. This popular testing method involves a heat map that shows where people click their mouses while looking at a page. Users then answer questions that shed light on the whys of their behavior. Google Analytics offers one form of this type of testing where they report the percentage of clicks on links and images within a page. A true heat map, however, will show in color (reds, oranges, and yellows) where people click the most. This gives you immediate and intuitive clues as to where user's attention is drawn while looking at your page.

How do you control attention? The short answer is to hire a well-trained graphic designer, because smart design can control eye-flow through a page layout. But with that do-it-yourself mentality, too often we see untrained marketing staff making updates to pages without considering smart persuasive design strategies.

We've arranged with the fine folks at HotJar, a service that offers a number of testing options and feedback mechanisms, to explain a bit more about strategies for using heat maps in the next section. After this section by Hotjar, we'll tell you about a powerful way we are helping one of our client's use the same system.

Provided by Hotjar with permission:

Optimize your Site the 'Hotjar Way'

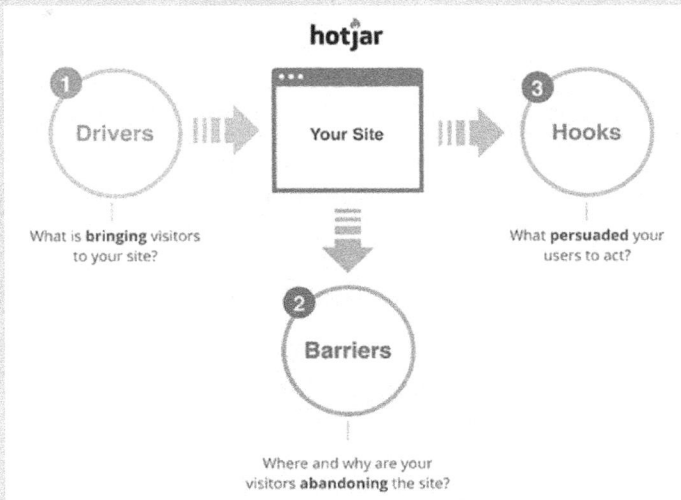

Drivers, Barriers and Hooks give you the 'Big Picture'.

Drivers - Discover your Visitors' intent.
Ask your visitors to use their words to describe what they are looking for – and why they want it – and you

will uncover powerful insights. Using the same exact words your visitors do will allow them to better connect with your site's content, interface and experience.

Understanding their intent will also make it easier to prioritize the right selling points higher on your pages. Finally – ask your visitors where exactly they found out about you. This is a great way to discover new and unexpected channels for growth.

Hotjar Tools that reveal Drivers: Polls, Surveys and Recruit User Testers.

Barriers - Uncover high blockage steps.

If you don't understand where and why your visitors are leaving your site you cannot really improve your site's experience and bottom line. So stop changing buttons and page layouts – and start focusing on breaking down your big barriers instead.

Always focus first on your biggest barrier. Your hottest opportunity is always the step or page with the highest traffic and the biggest drop off – so start there.

Hotjar Tools that uncover Barriers: heat maps, Visitor Playback, Conversion Funnels, Feedback Polls, Surveys and Recruit User Testers.

Hooks - Reveal the elements that are persuasive.

What are your customers really buying from you? Are they really buying a product or service or are they buying into a new lifestyle and status?

Understanding what really persuaded your existing users/customers to act or convert is the fastest way to converting even more of your visitors. It will also help you understand what will keep them coming back for more.

Hotjar Tools that reveal Hooks: Polls, Surveys and Recruit User Testers.

The 9-Step Hotjar Action Plan.

1. Set up a heat map on high traffic and high bounce landing pages.

2. Discover 'drivers' with feedback polls on high traffic landing pages.

3. Survey your existing users/customers via email.

4. Set up a funnel to identify your site's biggest barriers.

5. Set up feedback polls on barrier pages.

6. Set up heat maps on barrier pages.

7. Use visitor playback to replay sessions where visitors are exiting on barrier pages.

8. Recruit user testers to reveal drivers and observe barriers.

9. Reveal 'hooks' with a feedback poll on your success pages.

(You can read and learn more on heat map testing strategies in Hotjar's guide for uncovering hottest opportunities for growth)

Now, here's a very important example of how we are helping a corporate marketing client who makes software for the healthcare industry to use heat maps. The short story is that we have been working with this client for several years and they were recently acquired by a larger company, also in the healthcare B2B space. We've been working with them on their content marketing strategy and weekly creation work, with a strong focus on three target personas they wish to engage that work in hospitals: revenue cycle managers (which could include the CFO); the patient access manager; and the schedulers/registrars who usually work at the front check-in desks.

Below is a heat map click test result for one of their software landing pages:

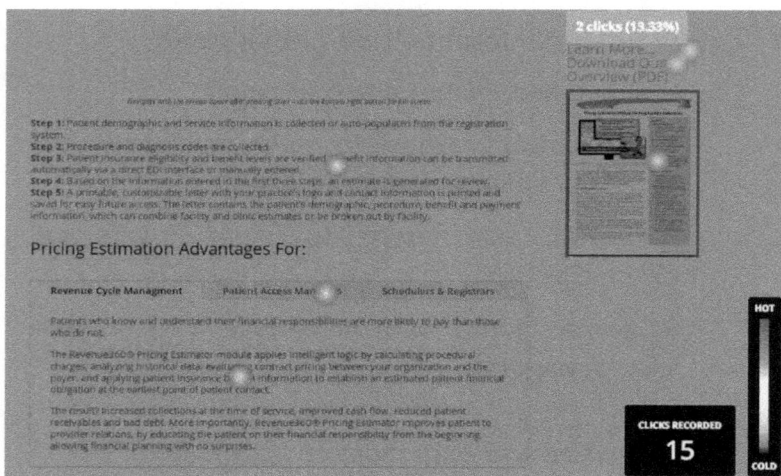

Above, the three tab sections at the left side of the page serve to engage and speak directly to each target persona. We're happy to already see "heat" on the second tab for patient access managers. Out of 15 page views so far, there are three clicks already to download the sales PDF at the right side of the page. This is a true conversion in which a page visitor moves one step closer to doing business with this client. The downloadable information sheets are important in B2B marketing where the decision cycle is longer than B2C and there are often several people that influence the final approval to purchase a product or service.

Why do we say this example is very important? Because now the marketing manager for the acquired company can show proof of conversions on landing pages thanks to content marketing efforts. And the new parent company is not even fielding a content marketing program or testing and tracking their own landing pages — at least not yet.

A/B tests

A simple A/B test involves rotating a different page with variations on the headline, graphic, and a CTA for every other visitor to your page. This will allow you to evaluate which page resulted in the desired outcomes you've established for your site. There are a number of programs, including one built into Google Analytics, that facilitate these tests.

Multivariate tests

A more extensive alternative to A/B testing is the multivariate test, which entails a rotation of different combinations of page elements so that several versions of a page are presented. For example, in a simple multivariate test consisting of two headlines, two compelling graphics, and two CTAs, users are presented with a total of nine possible versions of a page. Because this test allows for more combinations, there is a greater chance that there will be a version that stands out with optimal results. Multivariate tests often result in better optimization gains, but require more visitors to the page to get actionable results.

Nav flow tests

A crucial part of your website's user interface lies in its navigation. With a nav flow test, you can ask your testers how they would get from the home page to, say, the event sign-up page. Then you can provide wireframes and mockups for the testers to review. This is another type of testing that should be done before anything gets coded.

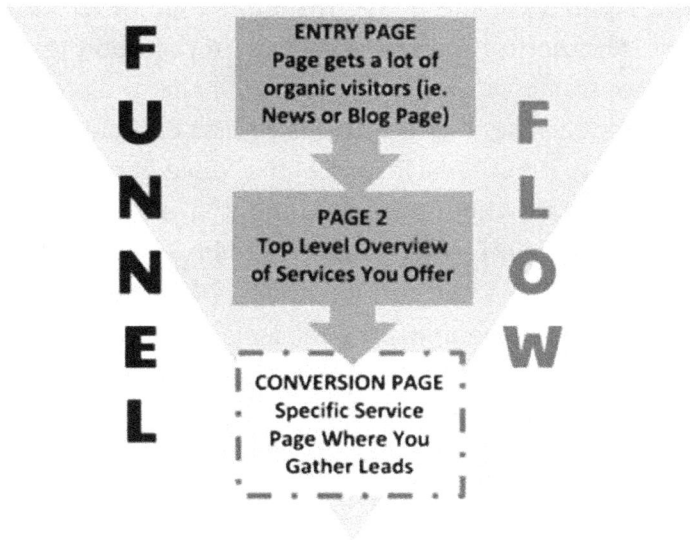

FUNNEL

FLOW

ENTRY PAGE
Page gets a lot of
organic visitors (ie.
News or Blog Page)

PAGE 2
Top Level Overview
of Services You Offer

CONVERSION PAGE
Specific Service
Page Where You
Gather Leads

Some marketers like to think of the way visitors flow toward a goal on their sites in terms of a funnel, as illustrated above. Optimizing visitor behavior to compel them toward a desired business outcome is a key concept in CRO.

Nav flow lets you upload a series of images that represent a path your users would take to get from one section to another. By identifying where users click on your navigation elements you can quickly figure out whether your navigation design and related CTAs are working effectively.

Recorded tests

Certain kinds of recorded tests offer insight into visitor behavior, including a cost-effective option that records a user's movements around the screen. This may not seem necessary given the availability of click analytics, but recorded movements can give you and your designers clues to behavior and page attention that other tests cannot. Another more expensive type of recorded test is to pay contracted reviewers to surf your site and provide verbal feedback as they go. This is similar to focus-group testing, though in this instance the reviewers do not meet and share their thoughts about a site together. Similar to the pre-screening of a film, this kind of recorded test provides key feedback while there's still time for last-minute adjustments. And now, with Skype and Google Hangout video chat capabilities, smart marketers can pull these kinds of reviews together much more easily.

Lead quality testing

In the end, businesses need sales, and for most business-to-business (B2B) and non-eCommerce sites, this means gathering not just any leads, but quality leads. This is often a manual process whereby your sales team might simply sort the leads you gather through forms on your website into hot, warm, and cold categories. You will have to take a closer look to determine what makes for the best leads, and what specific qualities you don't want. For example, you may be getting leads from people who have no budget for your product or services. If that's the case, then one simple adjustment would be to add a pop-down qualifying field on your lead forms that asks for their available budget numbers. Good sales teams will have some kind of lead scoring in play, using their customer relationship management (CRM) software to sort leads into levels of interest, buying intent, and other factors. Sales needs to give feedback to the marketing and web development teams so they can work to ensure that higher quality leads are brought into the lead nurturing process.

Verbal feedback tests

In this kind of test, you solicit feedback after people review a design element or brand messaging. For example, you can ask which version of a logo would best represent a certain company, or which stock image would be most engaging for a specific event.

Exit surveys & polls

Another way to hear what real customers think (as opposed to what execs think) is to do an exit survey. One good solution is the 4Q Framework by iPerception. There are many survey plugins and methods available, but 4Q offers a robust community that will help you analyze your results. The 4Q—or four questions—are designed to identify who is at your site, why they are there, how you are doing for them, and what you need to fix. You can also use your own forms program and a pop-up plugin that triggers upon leaving a site; an example of one in play can be found at our CollegeOfMarketingPros.org (COMP) site.

3

What needs to be optimized at your current site?

You are not optimizing just a page or a site. Rather, you are optimizing for desired visitor behavior.

Chapter 3. What needs to be optimized at your current site?

There are a lot of TLAs—three-letter acronyms—in this business. When it comes to optimizing at your current site, a lot of people think it is either keywords, commonly referred to as search engine optimization (SEO), or landing pages, otherwise known as landing page optimization (LPO), that you must consider. While some of that is true, here's a newer, big-picture TLA for you: VBO, or visitor behavior optimization. You see, a lot of people are focused on details, semantics, and design elements to the point where they forget that they are not optimizing a website, but rather, optimizing for what they want people to do at a website. Here are some key areas to consider:

Usability

Can users easily, intuitively, and efficiently perform the tasks you want them to at your website? Often this has to do with the clarity of your site navigation, which is a vast enough topic to easily be the subject of several chapters in a book about user interface design! Here we'll just present three tested and proven recommendations:

1. **No more than seven items on top-level navigation**. An individual's short-term memory can hold only seven items, thus it is easier to assimilate and comprehend what a site is about when people are not presented with more than seven navigation items.

2. **No content more than two clicks away from the home page**. In other words, make it easy and fast to find your content.

3. **Make site content and navigation intuitive with visual clues**. The use of icons, backgrounds art, colors for different sections, and more subtle cues from the discipline of graphic design and layout will help you here. Did you know you can guide eyeflow using graphic design elements? Oftentimes the best practices will not be conscious to your visitors, but site users will find what they need easier, and they will likely take note of your site's user-friendly design.

Let's finish this thought with an analogy about finding your way in a large museum. Most museums are well organized and present visitors with maps to displays and exhibits. This way you know exactly how to quickly reach what you want to see. In Portland, Oregon, near the *Webdirexion* headquarters, there is a famous bookstore—*Powell's Books*—that spans over a city block and has several floors filled with floor-to-ceiling shelves of books. Where does one begin to find the books they seek? Well, the staff worked this out, in part by naming different rooms with a color system so that the "Purple

Room" might house philosophy books, and the "Red Room" might house travel books. This is exactly how you should think about organizing your website to optimize it for usability.

Engagement

Are users engaging with your content? Some important analytics benchmarks for this are bounce rate (when a user leaves a page) and page views per session (the number of pages a visitor to your site looks at before leaving). Note that three or more page views per session generally denotes an "engaged visitor," because studies have shown that new visitors who look at three or more pages on your site are on average about 10–25 percent more likely to do business with you.

A word of caution: Do not obsess over bounce rates and similar stats. Consumer behavior and psychology are complex, and you are only seeing part of the picture with analytics—what site visitors are doing and what they are not doing. Can you optimize for improvement? Absolutely. One way to keep people moving through your site is to use analytics to identify the most popular content and then place prominent links to such content above the virtual "fold," which refers to the top of the screen before a user must scroll to see what is farther down on the page. But you do not want to worry about bounce rates too much. Why? Again, they do not paint the whole complex picture of what visitors do on your site and why they do it. For example, if you have an interesting blog, it is not uncommon for a certain percentage of visitors to come to one new blog post then leave. This would show up as a bounce rate in Google Analytics, but it does not mean that you have failed to engage visitors in a good way.

Relevance

This focus is of paramount importance to Google, as well as to SEO. But it shouldn't take Google to tell you that your site must be relevant to your visitors. You won't sell anything if visitors cannot find pertinent information and related products and services. And if your content does not help visitors satisfy a need or want, then kiss them goodbye. So how do you optimize for relevance? Start by identifying what is important to the visitors you intend to engage and then serve them what they need. Remember that it's always a good idea to ask them. So many times we see companies with a high-level executive dictating what to write, and how to phrase things—even what keywords to use for SEO. This is inside-out thinking, and it is very likely that what is generated in this manner will be irrelevant to visitors.

Instead, deploy outside-in thinking where you go outside your organization to study facts about your target market, including what search terms they use, what sites they visit, and what their demographics might be. Learn their pain points, the key motivational factors that have them searching in the first place. Then, and only then, you can begin to create content that addresses real-world motivations and the intent of your target personas. The Internet provides great opportunities to more easily study the needs and intent of your target groups, and those who take the time to consider this will have more success in optimizing their sites for relevance.

How do you optimize for the quality of leads? Recall that the term qualified lead was mentioned above in the section giving a formula for valuing your leads. The first assumption here is that your marketing team will actually talk to your sales team and ask them, "Are we delivering high quality, closeable leads from our website, or are you getting too many tire kickers who are not interesting in buying?" Once you know this—and good teams with a smart CRM solution will actually score their leads appropriately—you can consider some ways to make optimizations at your site to attract and engage more qualified leads. Here are two suggestions:

1. Make sure you have content that is relevant to different phases in the buyer's journey. Is your visitor just becoming aware they have a need, actively seeking options, or ready for a proposal? Prepare content that addresses the psychology of different potential buyers who are in each of these key phases of their hunt for solutions.

2. Use lead form questions to help you optimize for more qualified leads. Ask questions about the prospect's budget, and how soon they intend to buy.

Contact Webdirexion

Tell us a bit about your company and marketing campaign goals...

Name *	[🛈]
Company *	[ ]
Email *	[ ]
Phone Number	[ ]
Website	[ ]
Our Company's Size Is: *	[10-20 Employees ⬍]
Our Annual Marketing Budget is: *	[$24k - $30k ⬍]
Please name your top three competitors and provide site address:	[ ]
The primary business objective of our site is (tell us what you want to measure for ROI):	[ ]
We are interested in: *	[A Mobile Site ⬍]
We are ordering an AUDIT/Site Fix Report/Consultation (optional):	[⬍]
Comments:	[ ]

Send to Webdirexion

This is our Webdirexion *lead form — note that our question about the annual budget of prospects starts at $24k–$30k. It's one thing to get a lot leads, but another to get qualified leads. Your marketing team needs to hear back from sales people if those filling out forms are really qualified to do business with you. Leads that don't have the right budget your services or products require are not qualified leads.*

4

Testing Before You Code a Page or Site

*A landing page or website design is simply an **untested** hypothesis, and visitors seldom react to a page or site the way designers think they will.*

You'll save time and budget costs by testing designs up front.

Chapter 4. Testing Before You Code a Page or Site

Many content marketers overlook the opportunity to test designs and encourage desired business outcomes. The key need here is to understand what your visitors want (their intent) and then evaluate your options in terms of interface elements. When you do this before you code your site, you will save a lot in programming costs. Priced out a good programming team lately?

For a recent interface test prior to coding, our client Joe Hage, CEO of *Medical MarCom*, collaborated with us to test a site we developed for *The 10x Medical Device Conference* before it went live.

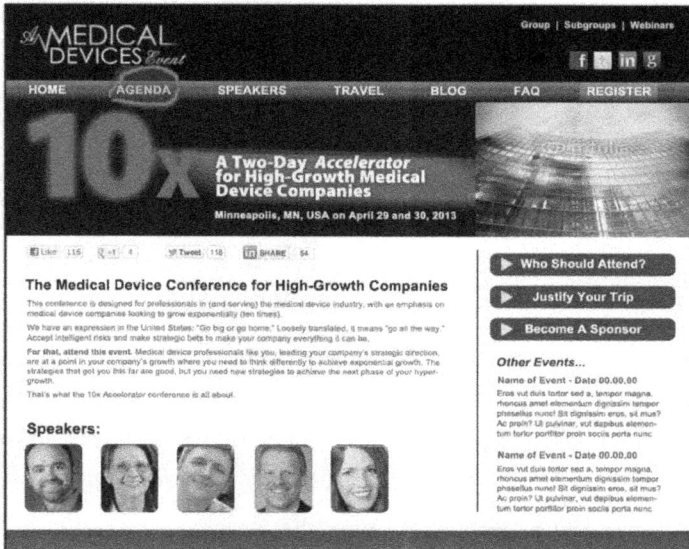

In this test, we had guessed that speakers would be more important to people; it turns out that the agenda was primary.

Above is one of four tests we did for *The 10x Medical Device Conference* site, before the design was coded. It's a heat map that shows where people clicked when asked what section would interest them most. We predicted that more people would be interested in clicking to learn more about speakers, but discovered that more visitors were actually interested in learning about the agenda, and also that some clicked on an image even though there was no clear CTA.

We solved a number of communication issues prior to coding, and the resulting changes we made quickly paid off in event registrations for the organization's first annual conference, based on the following principles:

- **Testing before coding saves money** — Smart content marketers review analytics and then refine how their site interface works to compel visitors to complete goals like downloading case studies and filling out lead forms. You pay a programmer or content administrator to make refinements, but why not learn what you need to do before you even code a new site? You'll be money ahead.

- **Make smart use of every click** — We learned that people were clicking on a static banner even though there was no indication of a CTA. So we linked it to the most popular content, which was also revealed during testing.

- **Design and marketing communication are both art and science** — Sure, we collected stats from our test (the science part) but we weren't afraid to go with our gut when it came to choosing a good image for the conference. Keep in mind that you can ask for written responses from test takers, too.

- **Go with the flow** — Visitors flow through your site on a conversion funnel path toward desired business goals. Testing reveals exactly how they want to flow up front. Then you can review analytics to double-check this flow.

Think through your test questions very carefully. You need to give a brief context statement (i.e., you are online surfing for medical device conferences), and then be precise about what you ask them to do.

We also tested *The 10x Medical Device Conference's* new design one year later (heat map test shown below). We posed this question with a qualifying context statement: "You're in the medical device industry considering events to attend. Assuming you watch the video, which CTA at right of video would most compel you to click? (click one)."

Here are four different CTAs at the right of the same introductory video. Which one do you think was more important for prospective attendees at a medical device conference?

Which of the four CTAs shown above triumphed above the rest? As it turns out, the first and second were more important to the testers than the third and fourth options. And since we ran another couple of tests indicating that the agenda reveal was primary in motivating interest, we ran with that. We'd like to give a hat-tip here to Joe Hage, a great marketer and the force behind the medical device event—the popular *10x Medical Device Conference*—we were working on in the above example. He's iterative in his approach to testing, looking for small gains on each round of study, and he also had the creative impetus to use the testing service FiveSecondTest.com outside the normal box of simply studying a web page user interface.

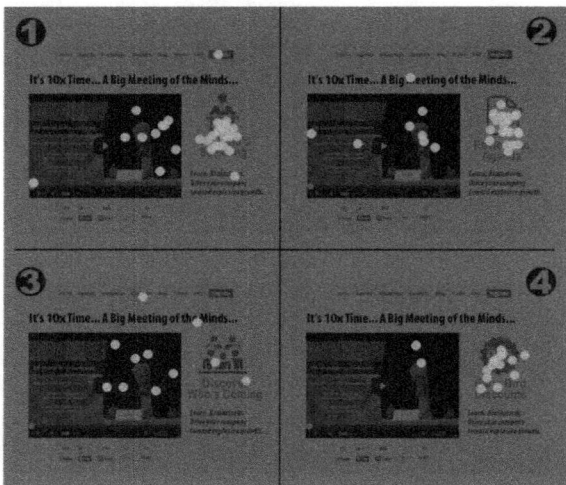

Here you can see how the resulting click map showed that #1 and #2 were preferred to those on the bottom row.

There were four CTAs, and visitor interest in who was speaking won out—even over a CTA for early-bird discounts. Implementation based on these results increased site engagement and proved useful when we refined the registration page with testing to increase conference registrations. The result? Close to a 50 percent lift in registrations over the same period last year. One last bit of credit to Joe Hage— he made sure that the testers were actual industry insiders so he could get a true read from them, and also do a bit of pre-conference promotion via the early testing.

One more story from the trenches: Throughout the years, we have worked on a number of websites in the *Content Marketing Institute (CMI)* family of properties, including the main website itself. For a redesign, we used five-second testing to show a JPEG of the new homepage and asked four questions. Here are the questions we asked along with summaries of the responses.

1. **What is the main purpose of this site?** A majority of respondents were able to quickly and correctly identify that the site was about content marketing.

2. **Would you sign up for the newsletter?** The answers here were split 50/50, with a number of users indicating that they did not see a sign-up form for the newsletter. This was an important conversion goal for CEO Joe Pulizzi, and one of the secrets to his ongoing success.

3. **If you would not sign up for the email newsletter, why not?** This was a follow-up question to learn what we could do to adjust for more conversions in the form of newsletter sign-ups. The majority of testers said that they did not see the sign-up box.

4. **How could you get helpful articles from this site?**
We threw that one into the mix because we had a theory that the online articles section was far from obvious. About 40 percent answered that they did not know how they could get articles from the site.

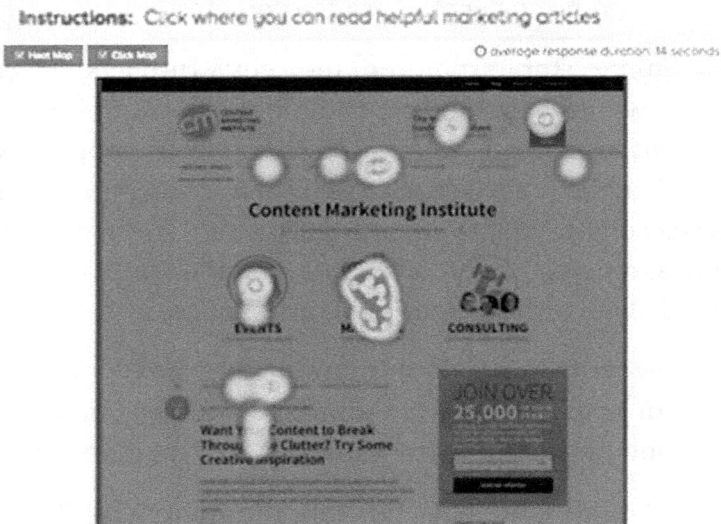

Above, CMI's homepage did not originally have a link to a homepage for their online articles. People clicked mostly on the print magazine icon in the center.

Before we coded the site, we realized that both the newsletter sign-up and the link to get online articles had to become more visible. *CMI* was trumpeting three new initiatives: the first-ever Content Marketing World event, the launch of a new print magazine, and new consulting services on the company's homepage. These three important elements had caused the web builders to miss making their own online articles obvious in the preliminary design.

The user interface was simple: a link to "articles" was added to the top-level navigation. Obvious in hindsight? Yep. Obvious in the heat of the battle? It rarely is. Now visitors can easily engage with a series of articles on content marketing and then be asked to sign up for articles to email via a pop-up (one solution to making the newsletter CTA more compelling).

5

SEO without CRO is like a Plane Without an Engine

SEO is an "inbound marketing" tactic that can help you get found and direct the right prospects, or travelers to your airport — to follow our analogy. Once they are at the airport, how do you get them to the right terminal, and then to their all important final connection? CRO.

Chapter 5. SEO without CRO is like a Plane Without an Engine

What is it that you want to achieve with your company's website? Is the end goal the sheer glory of seeing your site appear high in the search engine pages for random keyword phrases? Or perhaps it's the result of C-suite executives' push for numbers—the more people that arrive at the site each month, the better?

Of course not. Search engine rankings and visitor stats are meaningless unless they convert into leads or clients. Even if the only goal of a website is to provide information, visitors who have an interest in the topic should be able to find the information they seek quickly and efficiently.

The point of a website is to get visitors to interact with it in one or more of any number of pre-determined conversion goals. SEO gets the visitors to the site, but CRO gets them to hit those conversion goals.

The objective of SEO

To take the above a step further, what is the objective of SEO? Is it to achieve high placement in the search engine results pages, or is it to achieve relevant site traffic?

If all you want is traffic for traffic's sake then, by all means, pepper every page with key phrases containing the word "free" and any number of X-rated buzz words—but don't expect your website visitors to do anything other than suck up your bandwidth.

SEO for SEO's sake is like a plane without an engine. You can fill it with as many people as there are seats, but you aren't going anywhere.

Avoiding ego boost keyphrases

Unfortunately, when it comes to digital marketing, some business owners just lose their minds. Rather than create content that provides the most relevant information to their target customer, they start pushing to get ego boost phrases to the top of search engine results pages (SERPs). At that point, SERP domination takes precedence over client conversion.

Seriously, who cares if your website is on the first page of a Google search for the term "general contractor"? Do you think that a motivated buyer is going to search with a vague term, or do you think he or she is more likely to use a location-specific term, such as, "Clark County general contractor quotes."

By getting more specific with your content, you are immediately reducing the competition for airspace at the top of the SERPs, and Google recognizes that you are far more likely to provide a relevant answer to any search query related to general contractor quotes in Clark

County (whether or not the searcher included the words "Clark County" or was simply located in Clark County).

Okay, so you can improve your SEO by using more specific phrases and, if relevant to your business, including location focused content. But this is not exactly news. Google has been geotargeting for several years now and we have covered choosing the perfect long tail keywords in past articles on the *Webdirexion* blog.

Good SEO, which includes things such as choosing the most effective long-tail keywords, can result in vastly improved traffic, but remember that high visitor numbers are not the end goal. It is what these visitors do on a site that matters, and that is where CRO comes in.

SEO from a CRO POV

Wow! Acronym overload! Let's forget how SEO can influence your SERPs and look instead at how it might influence your visitors and increase your conversion rate. If you optimize your site with well-chosen keywords, then visitors that arrive via a Google search listing will find what they are looking for. On the other hand, if you have gone for ego boost or ineffective phrases then you will see high bounce and exit rates.

In Google Analytics, you can view bounce rate vs. sessions for the entire site, or drill down and analyze bounce rates for a particular page over time.

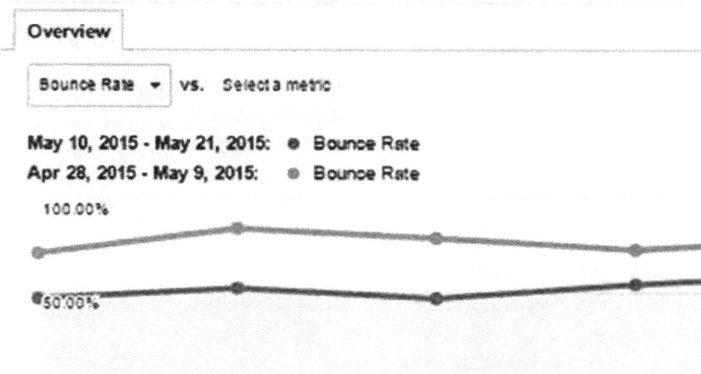

Overview

Bounce Rate ▼ vs. Select a metric

May 10, 2015 - May 21, 2015: ● Bounce Rate
Apr 28, 2015 - May 9, 2015: ● Bounce Rate

100.00%

50.00%

You can also identify which pages have the highest exit rates and flag them for review. The Google Analytics behavior flow chart (example below) allows you to see which pages have the largest drop off, and you can view this by landing page or by medium/source. This will help you determine how tweaking site content/keyword selection is affecting your visitor's path through the site.

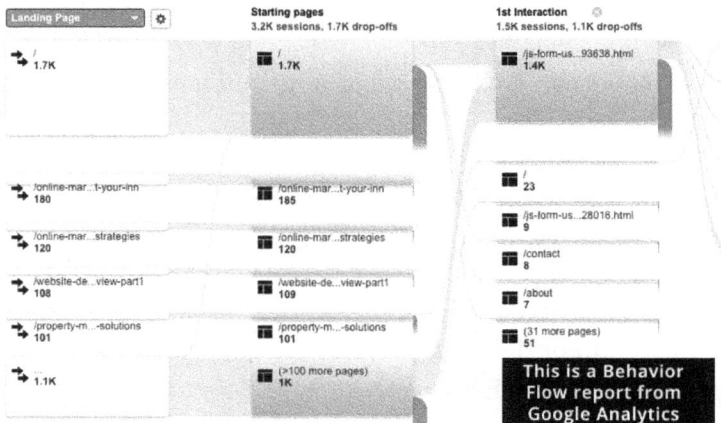

This is a Behavior Flow report from Google Analytics

With your Google Analytics dashboard you can access reports like these any time, but it's possible to set up a specific content conversion report. A good starting point is the Content Analysis Dashboard template by Vagelis Varfis, available in the Google Analytics Solutions Gallery - https://www.google.com/analytics/gallery/#landing/start/

III ✪ ⚑ Content Analysis Dashboard

By Vagelis Varfis | Nudge Digital   Mar 14, 2014

★★★★★ (24) ➕ 23,990 🏴 ⌨ **8+1** **f Like** **🐦 Tweet** ✉

This Content Analysis Dashboard is all about analysing and providing insightful data that will help you evaluate the efficiency of the content in your website. Based on these widgets you will be able to see which one of the pages are underperforming/overper...

More by Vagelis
Varfis | Nudge
Digital

[Import]

Unpack that template within the Custom Report section of your site's Google Analytics account and you'll have a dashboard that shows you:

- Page views and unique page views by page title
- Visits and percentage of new visits by landing page
- Average time on page and bounce rate by page title
- Exit and page views by page
- Goal conversions
- Page views by country/territory
- Page views by city

You can add or delete up to 12 widgets, which allows you to slice and dice visitor activity in just about any way you want.

Reports can be set to email interested parties on a regular basis and it's a good way to keep your SEO team focused on its CRO goals: converting visitors into customers.

6

Setting Measurable Business Outcome Goals

How will you know if you've been successful unless you set clear objectives to be met?

Remember, you can't take credit for what you haven't measured.

Focus on S.M.A.R.T. Goals: Specific. Measurable. Attainable. Relevant. Time Bound.

Chapter 6. Setting Measurable Business Outcome Goals

Having desired business outcome goals at your website and landing pages (and knowing the value of a conversion goal) is arguably the most important part of any CRO program. Everything else depends on this. If you don't know what you want visitors to do when they get to your site or a landing page, then you can never make any improvements. And you need to value what they do with real-world dollar figures as close to sales-related income as possible. This could be the subject of a couple of chapters, but to boil it down for our CRO guide, we'll look at two key factors in goal success: common website goals and how to value them.

Common website goals

A good starting point is to focus on three types of goals—the things you want your site visitors to do. You want them to engage with three or more pages of content; you want them to download a key PDF or two; and you want them to sign up for a newsletter or ask for more information about your products and services.

Formula for valuing goals

Now that you have your initial goals, how do you value them in a meaningful way? Start with the general income earned by a sale. In B2B, it is not uncommon to have a high-value service product that might cost $1000 or more. For the sake of example, let's say your company makes a $1000 sale. Next you need to know your sales team's close rate. In other words, for every 10 quality leads your sales team gets, what percentage will become customers? Let's say the answer is 30 percent, meaning they average three sales for every 10 leads. Ten leads are thus worth $3000 to your company, or $300 a piece. So, you would enter $300 in analytics as the value of a qualified lead at your site.

We have a worksheet to help you calculate goals, in the back of this guide.

Goal description   Edit

Name: Audit Lead
Goal type: Destination

Goal details

Destination

Always create a custom
thank you page.

Equals to ▾   http://Webdirexion.com/thank-you   ☐ Case sensitive

For example, use *My Screen* for an app and */thankyou.html* instead of *www.example.com/thankyou.html* for a web page

Value OPTIONAL

This dollar goal value will be factored into every
indicator being tracked.

On   300.00 $USD

Assign a monetary value to the conversion. *Learn more about Goal Values.* For a transaction, leave this blank and use Ecommerce tracking and reports to see Revenue. *Learn more about Ecommerce Transactions.*

Funnel OPTIONAL

It's a good idea to track flow through your
site to a measurable goal.

On

Use an app screen name, the string or a web page URL for each step. For example, use *My Screen* for an app and */thankyou.html* instead of *www.example.com/thankyou.html* for a web page.

Step	Name	Screen/Page	Required?
1	Home	http://webdirexion.com	No
2	Audits	http://webdirexion.com/may-audit-special	

+ Add another Step

As illustrated in the Google Analytics image above, we entered $300 as the goal value for a lead in Google Analytics. We recommend a custom thank-you page, not only for tracking the conversion (visitors only see that page if they fill out a form or take some other conversion action), but because you can then redirect and suggest other review items on that page via links after you thank them.

Why is it so important to enter and follow dollar values for goals in analytics? Because a good analytics program such as Google Analytics will track every single interaction on your site and assign a dollar value to each based on a set of predetermined goals.

Social Network	Conversions	↓ Conversion Value
	45	$1,350.00
	% of Total: 3.58% (1,256)	% of Total: 3.93% ($34,325.00)
1. LinkedIn	24 (53.33%)	$750.00 (55.56%)
2. Twitter	13 (28.89%)	$340.00 (25.19%)
3. Pinterest	3 (6.67%)	$90.00 (6.67%)
4. Facebook	2 (4.44%)	$50.00 (3.70%)
5. Google+	2 (4.44%)	$80.00 (5.93%)
6. Disqus	1 (2.22%)	$40.00 (2.96%)

As illustrated above, setting dollar value goals in Google Analytics allows us to see the total value of visitors from each social network.

Setting dollar values for goals will help you compare success of landing pages:

Landing Page ?	Acquisition		Conversions Goal 1: Engaged Visitor ▼		
	Sessions ? ↓	% New Sessions ?	Engaged Visitor (Goal 1 Conversion Rate) ?	Engaged Visitor (Goal 1 Completions) ?	Engaged Visitor (Goal 1 Value) ?
	8,455 % of Total: 100.00% (8,455)	80.27% Avg for View: 80.28% (0.01%)	8.37% Avg for View: 8.37% (0.00%)	708 % of Total: 100.00% (708)	$28,320.00 % of Total: 100.00% ($28,320.00)
1. /	5,069 (59.95%)	80.53%	5.48%	278 (39.27%)	$11,120.00 (39.27%)
2. /wordpress-support-service-plans	318 (3.76%)	88.36%	1.57%	5 (0.71%)	$200.00 (0.71%)
3. /online-marketing/four-marketing-funnel-strategies	295 (3.49%)	87.12%	16.95%	50 (7.06%)	$2,000.00 (7.06%)
4. /inbound-marketing/online-healthcare-marketing-mayo-clinic	246 (2.91%)	91.06%	15.85%	39 (5.51%)	$1,560.00 (5.51%)
5. /inbound-marketing/social-media-marketing/social-media-marketing-content-strategy?utm_hashtag=#Rev360	227 (2.68%)	90.75%	9.25%	21 (2.97%)	$840.00 (2.97%)
6. /website-design-and-development/wordpress-site-design/understanding-wordpress-from-a-beginners-point-of-view-part1	190 (2.25%)	24.21%	5.79%	11 (1.55%)	$440.00 (1.55%)
7. /online-marketing/healthcare-marketing	138 (1.63%)	92.75%	26.09%	36 (5.08%)	$1,440.00 (5.08%)

We can now view the dollar value for a specific time period for each of our landing pages and determine the relative worth of each. This can be helpful when there are meaningful goals set that would apply to more than one of the pages.

You can't take credit for things you don't measure. We'll conclude with some questions that will help you think about what to measure. These questions come courtesy of an article by Shelley Koenig at ConversionScientist.com (used with permission):

- Will "new" visitors behave differently than "returning" visitors?
- Do you have a large number of mobile visitors? if so, is their RPV — Revenue Per Visit* — higher or lower than desktop visitors?
- Does PPC (pay per click) traffic have a higher RPV than organic traffic?
- Does email traffic have a lower abandonment rate?

*RPV is a key performance indicator (KPI — the acronyms march on) most often used on eCommerce sites. For B2B services sites, you can still use analytics to show value during visits following the guidance in this chapter.

7

Testing Ads: In Sync With Landing Pages

In order to achieve a high quality score from Google when buying ads on their system, your landing page must synchronize well with your ads. Then you will achieve better ad positions, and better pricing for your ads.
Why?
Google lives or dies depending on one word:
Relevance.
Google wants your landing pages to be highly relevant to your ad copy. Relevance sells.

Chapter 7. Testing Ads: In Sync With Landing Pages

When it comes to online advertising—or even old-school advertising, for that matter—testing always pays off. And nowadays, technology makes it easy to test two headlines in a text ad—or rotate in two different graphic designs for a display ad—and learn which versions visitors respond to best. Back in the day, it was harder to do such tests because the creative process took much longer without the use of digital production methods. But even then, advertising legend David Ogilvy admonished:

> The most important word in the vocabulary of advertising is 'test.' Test your promise. Test your media. Test your headlines and your illustrations. Test the size of your advertisements. Test your frequency. Test your level of expenditure. Test your commercials. Never stop testing, and your advertising will never stop improving.

So, the principle and need for testing online ads is the same, but the intent and motivation of the target audience is slightly different. On a website, you are testing to compel an outcome that engages visitors with your business, using tools such as email newsletter sign-ups, etc. With an advertisement, however, you are simply testing to pique the searcher's curiosity, one step back from a business engagement outcome. You want a click-through, which is not quite a signal of interest in a specific product or service. Your ad may mention a specific widget you are selling, but you are not yet trying to get the visitor to agree to be contacted by your sales team.

The many facets of a Google Quality Score.

Landing pages should usually work in tandem with online advertising in more ways than one: Google and other online advertising services will score how well ads match the relevance of pages and you need a good deal of targeted traffic to get a scientifically valid tally for which page version pulls better results. Even when ads lead to a more standard homepage, the ad should help indicate to visitors what you will ask them to do at the website. Done right and it's a left hook followed by a roundhouse punch, a knockout 1-2 sequence. Done poorly, the visitor will not understand how your landing page expands on your advertising message and he or she will simply bounce away!

Here are some major focus areas for testing ads:

- **How many ad versions to test?** We always test at least two versions of a text ad or visual display ad, and often three (with as many as five) versions tested over time. You'll learn more about your visitors as your campaigns proceed, and thus be able to adjust ads for catch phrases that resonate as you go. After about four months, you may have narrowed down the ads to just a couple, but it is always a good idea to test another variation or two every month. Oftentimes, even changing a single word in a headline can boost conversions.

- **Google Quality Score depends on landing page synchronization.** Q stands for quality, and Google judges the quality of your online ads along with your landing page based on relevance and consistency to create its "Q Score". This, in turn, is used by Google to determine your ad rank—where your ad will appear on a search return page—so it's very important to get this right. Another reason to tune into Q scores? They're used to determine the price you pay: the higher your quality score, the less you will pay. And how do we keep improving Q scores? Testing.

- **Test your ads and test your landing pages.**

Sta　Keyword: **restaurants**

Showing ads right now?

No ⟩ • Your keyword isn't triggering ads to appear on Google right now due to a low Ad Rank. Ads are ranked based on your bid and Quality Score. What can I do?

• One of your other ads is already showing for this keyword. What can I do?

firs
bid　Quality score Learn more
First
bid　3/10 ⟩ Expected clickthrough rate: **Below average**
estii　　Ad relevance: **Below average**
€1.£　　Landing page experience: Average

The above is an example of a quality score with below-average results. Note that Google has reviewed the associated landing page and deemed it "average." A 3/10 score would cause the associated ad to be both more expensive and displayed less frequently.

- **Qualifying leads:** Are you getting leads that don't have the right budget for your products and services? This can be a good time to test using pricing for services in an advertisement. You can do so and still provide incentive, such as offering discounts or add-on freebies.

- **Creative approaches and design tactics:** If you intend to use visual display ads in your campaign, it will be very important that you work with a skilled graphic designer. Why? Well, even though there are many online programs that can help you throw together an image and text, a true ad-experienced designer can control eye flow in your ads. Yes, you read that right. Experienced professional designers can control where your viewers look in an advertisement. They can also influence moods and appeal to emotions. Advertising studies show that people buy into emotion, not rational bullet points. Get professional design help for better results, then test at least two versions of your display ads to find out which ones score highest. Note that you can also put up the two versions for five-second tests in order to get pre-deployment feedback. The environment won't be quite the same as when your ads will actually run (simply because active searchers have different mindsets than test reviewers), but it can get you some preliminary feedback that will help you perfect your ad versions before you start paying for clicks.

8

Smart Thinking Behind Smart Forms to Boost Conversions

Most marketers place at least one simple contact form on their sites. However, forms have become much smarter in what they can do to gather information and assist visitors. Think of them as robotic helpers and you will be way ahead of the competition.

Chapter 8. Smart Thinking Behind Smart Forms to Boost Conversions

There are a number of steps you can take to impact how many site visitors will convert to leads. Developing and testing forms is one tactic marketers often overlook. One approach is to create a smart form that delivers specific information tailored to the individual completing the form. Forms that automatically show or hide fields based on the user's response are known as smart forms, and FormStack refers to this capability as conditional logic.

Using this feature offers many advantages, including the following:

- Enables the form to look less intimidating

- Creates a unique, personalized experience

- Hides selected information that you don't want all users to see

FormStack released an extensive report that evaluated trends, conversions, and analytics of more than 40,000 forms for 2014. It's no surprise that smart forms came out on top, with the most engagement and conversions.

Using conditional logic

We recently created a church membership sign-up form for a client by taking the existing paper version of the form and making it digital.

St. Malachi Parish
2459 Washington Ave.
Cleveland, Ohio 44113

Confidential Parish Census- Personal information will be held in confidence

Digital form keeps data neat and organized and can be integrated to feed directly to contact mangament program

Census information can be cluttered and has to be sorted through.

| Paper Form Census Report | Online Census Sign up |

The form in question was extensive, allowing for up to seven family members to sign up at once. So we definitely had our work cut out for us. Three of the major issues we faced were:

- Breaking up the lengthy form to make it user-friendly and digestible

- Individualizing the form to display information that was only relevant to the person completing it

- Pleasing the client by still collecting all the important information

Simplifying an overwhelming form

Keeping forms as short as possible is a best practice,
but sometimes you really need extra data
in particular circumstances.
That's where conditional form logic is very handy.

–Jay Baer, founder of Convince and Convert

Using conditional logic not only allows you to shorten the steps taken for certain visitor personas, but also creates the opportunity to present different sales qualification messages based on earlier input they give in a form.

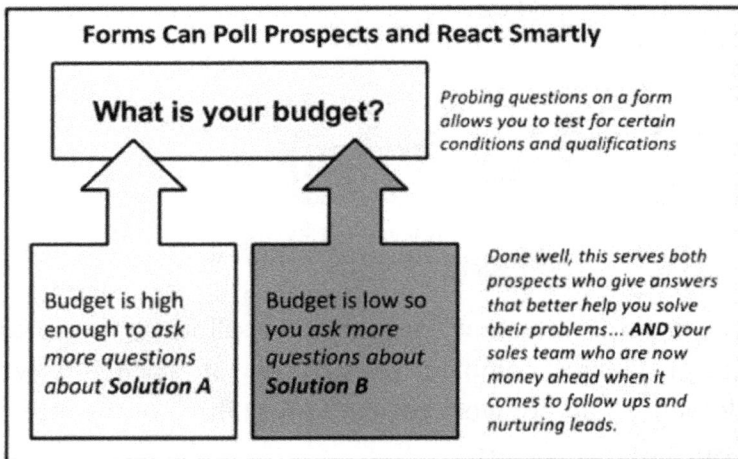

Forms Can Poll Prospects and React Smartly

What is your budget?

Probing questions on a form allows you to test for certain conditions and qualifications

Budget is high enough to *ask more questions* about **Solution A**

Budget is low so you *ask more questions about* **Solution B**

*Done well, this serves both prospects who give answers that better help you solve their problems... **AND** your sales team who are now money ahead when it comes to follow ups and nurturing leads.*

Using forms like the one shown above is an opportunity to poll your prospects and test for different qualifying factors. When visitors answer one way, question reponse A is displayed. When they answer another way, response B is displayed.

Are you signing up other family members?	Are you signing up other family members?	O Yes ● No
Do you want your family to be included in the printed Parish Directory?	Do you want your family to be included in the printed Parish Directory?	◉ Yes ◉ No
May we use photos of your family on our website (we may take photos or videos during liturgy and at parish	May we use photos of your family on our website (we may take photos or videos during liturgy and at parish	◉ Yes ◉ No

As the form we had to create for the church was very long and overwhelming, we decided to break it up into two pages in order to make the content easier to digest. According to *FormStack's* analysis, two-page forms actually yield higher conversion rates than their single-page counterparts.

Making the form relevant

As illustrated in the image to the right, the form we created was dynamic, changing depending on whether the person filling out the form opted to sign up multiple family members.

Hiding unnecessary fields

The church sign-up form can be used to sign up either an individual or an entire family, eliminating the need to create and maintain two separate forms.

We dedicated the entire second page to hold fields for additional family member information, so that while single-member users can submit from the first page, those signing up large families will be directed to a second page.

Maximize to get the most out of your form's performance. In the digital age, more and more companies are ditching the traditional pen and paper cards or registration forms in favor of streamlined online smart forms. We use a robust system called Contact123 Forms to create a versatile online form. These are included in our annual coverage plans and come standard for all retainer clients.

9

How to Increase Your Conversion Rates

Once you have resolved to establish a culture of testing, you will be on your way to gathering more qualified leads that your sales team can turn into customers.

Chapter 9. How to Increase Your Conversion Rates

CRO is an art and a science involving continuous testing to create more leads. What are your key conversion rates? How do you increase conversion rates? Answers to both questions are vital to your business, but a good number of marketers don't know their conversion numbers and aren't taking the right steps to increase results.

First, let us tell you what CRO is *not* about.

CRO is not about traffic volume. You could have a low volume of visitors in a very small niche, for example, but be converting a good number of those visitors into prospects that become customers.

CRO is not (primarily) about content marketing, because—as we said before—content is not king. Content is important in first engagement, but now you want to focus on another marketing c word: connection. Connection cannot happen until after you make conversions (another important c word). Too many marketers are so concerned with creating great content that they fail to focus on what happens after visitors engage with content. They overlook the importance of improving their rate of conversions leading to authentic connections that in turn grow sales.

In a nutshell, you need to test at each key point in your funnel for optimal results, from headlines in PPC ads to CTAs on landing pages to open rates in follow-up drip marketing. Then you'll win a higher percentage of qualified leads. It's as easy as that—and as hard as that. Huh? Well, the easy part is that a good number of tools are available to help you analyze conversion rate success. The harder part is to always be testing your marketing pieces for the best performers. It takes some time to make this a habit.

For now, here's a five-step CRO exercise for you:

1. Set measurable goals for conversions you can review in Google Analytics.

2. Look at your top landing pages and determine how they are performing to convert visitors to leads.

3. Determine what communication factors will better gather more qualified leads you can close (time to check in with your sales team, and review CRM reports).

4. Based on your initial analysis, determine what elements you want to test on top landing pages to convert more visitors to better-qualified leads—i.e., headline, graphic, CTA.

5. Start your first A/B tests for PPC ads, landing pages, and email subject lines. All three, ideally. Get the results, analyze them, and repeat.

10

CRO & Persuasion Psychology

All of the changes you need to make on a landing page, in an advertisement, in an email, and throughout your funnel flow should be based on one objective: persuasion.

"All of communication is an attempt to persuade."
— Dr. Stephen Ward, Communications Professor

Chapter 10. CRO & Persuasion Psychology

- In her book *eMarketing Strategies for the Complex Sale*, Ardath Albee writes about using personas and differentiating yourself with "attraction marketing."
- Tim Ash, author of *Landing Page Optimization, The Definitive Guide to Testing and Tuning for Conversions,* writes about empathy as a key ingredient to understanding your audience
- In *Influence*, his provocative book on the psychology of persuasion, Robert B. Cialdini likes to talk about social proof as one of the "weapons of influence."

We're going to weave those thoughts together in this chapter with a sprinkling of stories from the trenches.

But first, let's pause and consider something. You are not really optimizing for a landing page, an email newsletter, or a website. Rather, you're seeking to compel a certain behavior from a select target group. So, in the end, the optimization is performed on your visitors and prospects. That may sound subtle, but it's very important. So many times in our line of work we see the technical SEO worker mired in thinking about Google algorithms instead of considering how to best communicate to real live people.

Some of you may be familiar with Rand Fishkin, CEO of *Moz* (formerly *SEOMoz*). *Moz* has become a company influential in insights for "rankings, traffic, links, social media, content, and brand marketing." We use their analytics in a toolset along with other key performance indicators (KPIs) to monitor and measure CRO outcomes for our clients. So it was interesting to read in one of Rand's blog posts a few years back in which he confessed that he only recently realized that he was actually doing marketing—not SEO.

That's the main point of this chapter. In the end, it's not about CRO tools or tactics that win you the king we know as connection—it's about human motivation.

Let's look at some real-world examples.

Personas and the right triggers: This real-world sales example shows how finding the right trigger can pay off big time.

I once worked in sales at a major computer and office outfitter and was tasked with suggesting and closing sales for office chairs. There were a few basic chairs available in the showroom, which could be customized by special order. Around 60 different colors, patterns, and leathers were available.

Here's a story about my in-the-trenches testing:

> *My pitch for a while was simply to tell the decision makers—in general small business owners and managers—that "for only $20 extra, they could have their chairs covered in a variety of colors and styles to match their office decor." Wrong pitch. I sold very few.*
>
> *Fortunately, I could quickly do a test by simply calling a number of stores where the same special order chairs were flying out of the store. I found the pattern. I discovered the right trigger by comparing notes with the leaders. Then I became the top salesman for special order chairs in the district, moving from sales from around two to five per month to an average of 25 to 40 every month. How? I learned the true trigger was that most of the special-order fabrics lasted five to ten times longer for only $20 extra, and pitched*

that. I probably would have been fired if I hadn't tried that test.

This illustrates one of the psychological principles at the root of successful CRO: understand your target personas and what triggers them into action. Color choice was not valuable to small business people, but the concept of longer-lasting materials was.

Imagine if the corporate office had gotten wind of this story and used combined content marketing (power of story) tactics with CRO (A/B email subject line test, etc.) to get the word out to all the salespeople in the company?

Most smart optimizers use the concept of personas to consider the people they are talking to on their landing pages and in emails and PPC ads. You can write up a fictional character with whom you wish to do business. For our healthcare client, we used a persona called Pam (short for patient access manager). She's the one you might see when you are checking in and it is she who has to deal with all the trials and tribulations of patients who are nervous and worried. Better understanding her needs and motivations helps us create better landing pages for our users.

In his LPO book, Tim Ash suggests looking at behavioral styles as you consider creating elements for testing. This is a way to put your visitors into useful categories for thinking about their typical behaviors. For example, in a model popularized by Myers-Briggs, one classification for your target group might be a *perceiver* who prefers flexibility and an open-ended orientation to making decisions.

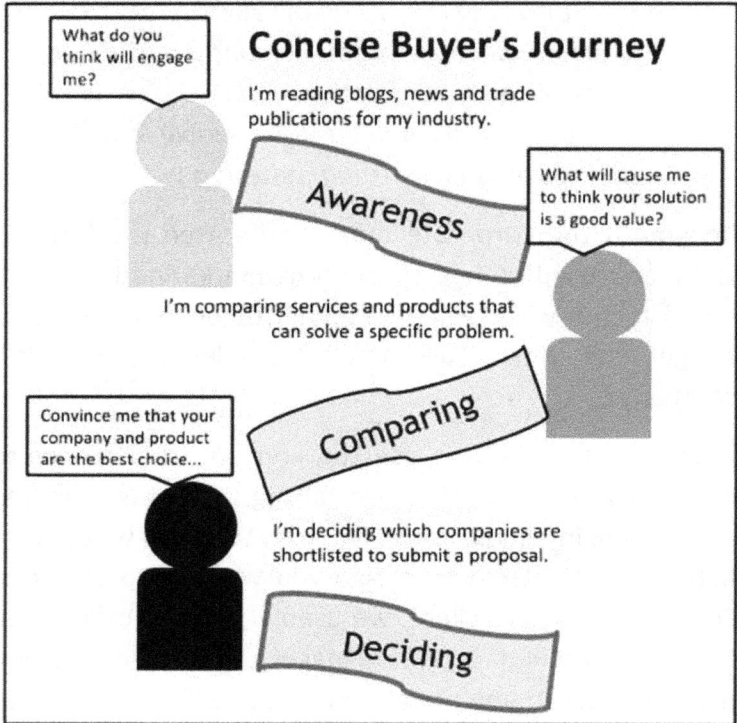

We find it very helpful to consider what phase in the buying process your prospect personas may be in when they reach your site and gear your persuasive components accordingly.

Yet another useful analogy, popularized by Joe Pulizzi and Robert Rose of *CMI*, at *HubSpot* and *Pardot*, is to structure your offers based on what stage of the buyer's journey your visitors are in. Are they just recognizing a need, comparing vendors, or nearing a purchase decision? Sometimes we will place both an informational

PDF (comparing vendors and doing research) and a CTA form (i.e., "Get a Demo Now"— for those nearing the purchase trigger point) on the same landing page to appeal to people in different phases of the journey.

Empathy, power of story, and LPO: *Webdirexion* has a niche focus in healthcare, with five clients in that industry. One of our teammates was an RN for six years before switching careers to become a writer and content marketer. For a client that sells software to hospitals, she used the power-of-story tactic (a content marketing technique where humans will actually put themselves into a setting and walk through a scene with the writer) to tell about how an old friend in the industry could really use some help from software to get through her workday as a hospital registrar.

The story allowed prospects—blog readers—to empathize with the protagonist. A good number of them followed through to a landing page where solutions to their pain points were provided and they could request a demo or download a PDF with more information. The demo was great for those closer to a decision point, with the informational download well-suited for those at an earlier stage in their buyers' journeys. We can also tactically decide to feature the story in the monthly eNewsletter and/or boost the post on LinkedIn to an appropriate target group.

To summarize, keep your buyer's journey points firmly in mind for CRO, flow, and LPO. Depending on what stage visitors are in they will be looking for different types of content at your site, and different CTAs will appeal to buyers in different stages. A buyer just becoming aware of options may wish to read your blog. A buyer comparing solutions will likely flow toward your service or product overview pages. And a buyer reaching a decision point will look for trials and demos.

Authentic social proof and other influencers

In Robert Cialdini's chapter on social proof in the bestselling book *Influence*, he details a series of innocuous examples, from bartenders salting their tip jars with dollars to start the process of proofs to laugh tracks and paid audience clappers and bravo callers (*claqueurs*, to use the professional term), both of which are techniques to get us to go along with the crowd. While Cialdini very honorably gives recommendations based on critical thinking about how to avoid the claptrap herd mentality, the major takeaway about social proof he demonstrates so vividly is that a significant number of people are guided by the behavior of others. This can be used in copy and landing page messaging the following ways:

- **Peer pressure** – Joe Pulizzi and the *Content Marketing Institute* team have successfully used the slogan, "Join over 140,000 of your peers" to go after that oh-so-valuable mailing list conversion. We remember when the number was closer to 20,000, but still compelling even then.
- **Testimony matters** – Get a good list of clients to review your products and services and add their testimonials to your landing pages. One tactic we have used is to trade recommendations on

LinkedIn, which then become public and may be used in other marketing materials.

- **Reviews get acted upon** – Savvy booking sites such as *Hotels.com* have reached out to industry clients with site widgets (small ads) stating their property has achieved, for example, an "outstanding rating of 4.6 based on over 129 verified reviews." The psychology? Over 100 people rated the venue highly, so it must be a good place to stay.

What about faking it? Buying traffic. Buying likes and paid endorsements. It should be obvious, but the answer is to resist this temptation. Will some dupes be initially persuaded when you appear to have some street credibility? Maybe. But none of the like-factory visitors will actually convert to customers, leaving you with only a questionable ethical conversion to the con-game scam mentality. People will sense when a company is propped up by false testimony and, in the end, the truth matters. People hate being conned, so the real value comes from heartfelt recommendations in the social realm. It can't be called "proof" otherwise, can it?

11

Case Study: Hotel Promotion Page A/B Test

The goal of a good landing page campaign is to lift results. What results? Usually it will be to gather leads via a smart form on your page as in the study we include here, but it can be to increase downloads of a case study, viewings of a video — you decide in each test what the desired business outcome will be.

Chapter 11.
Case Study: Hotel Promotion Page A/B Test—100% Lift

One of our clients is *Croc's Casino Resort* in Costa Rica, a large resort with a casino and three restaurants. We wanted to do an A/B test to improve conversions for one of the property's special promotions.

This is the original page: the A version for our test.

The resort was offering a three-night stay in one of their condo units for their grand opening in February of 2015. Our job was to increase visitor conversions (meaning the number of people who filled out the form shown on the right in the screen capture).

Next, as with any test, we had to decide which layout and content elements to vary, bearing in mind that the more elements we test, the harder it is to get a fix on which changes actually make the difference. If you intend to make a series of iterative changes, it is very important to isolate which element helped win the lift in conversions and which ones did not make a difference during the test.

While multivariate testing can help you test more iterations of landing pages, those sorts of tests are more time-consuming to set up and require more visitors in order to get enough data to make a decision. In the case of *Croc's*, we elected to focus on four elements and go for an increase in leads via one A/B test.

When focusing on a simpler A/B test method, we like to isolate and concentrate on three or four elements on a landing page. In this case we chose the headline, the graphic, the CTA, and the body copy. We probably could have focused only on two or three elements to save time, but in reviewing the page we saw ways we could strengthen all of the major page components.

Our testing element recommendations:

- If you test only one element, work on compelling word changes in the headline. Studies show that changing just one adjective can make a significant impact on your results.
- If it's possible to test two elements, we like to focus on either the headline and the CTA, or the graphic plus the CTA. Either of these combinations could be part of a second iteration test after you review your first A/B results.

Above are the elements we elected to test. As you will see, we also changed the background image in the B version.

When you are doing only one test, you might choose to address all the components of your landing page, as we did for *Croc's*. Your team should assess each component that comprises your landing page's case argument and look for persuasion enhancers in both graphics and copy. Recall that there is always some visitor reluctance to give contact information and that you will want to minimize potential triggers of friction (too long a form is one example of friction in the context of CRO).

B version (the winner)

Here's the B version of the page. We were fortunate to have a video to test in place of the original still image — the still image sold the "steak" (picture of Hotel), while the video helped sell the "sizzle" (what you can do while staying there). Note also that we began the top headline with a "magic word", "Win." Sometimes just a simple word change or two will increase your conversion results.

What we did: We added a video overview of the resort (it was done with renderings, because the resort was not yet built); added the term "VIP" in the main headline; highlighted some key points in the body copy; and added some qualifying details about viewing the condos above the form.

Here's how we reworded the second headline:

A Version Wording: Next 5 Condo Buyers in July Win a VIP Trip for Our Grand Opening Party

B Version Wording: Win a VIP Trip for OUR Grand Opening Party For The Next 5 Condo Buyers in JULY

"Magic" word.

The strategy here was to rewrite to let visitors know they had five chances to win more directly.

Results: 100% lift

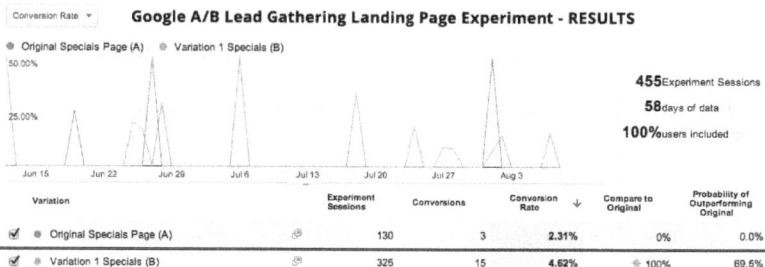

Conversion Rate ▾   **Google A/B Lead Gathering Landing Page Experiment - RESULTS**

● Original Specials Page (A)   ● Variation 1 Specials (B)

455 Experiment Sessions
58 days of data
100% users included

Variation		Experiment Sessions	Conversions	Conversion Rate ↓	Compare to Original	Probability of Outperforming Original
☑ ● Original Specials Page (A)	🗁	130	3	2.31%	0%	0.0%
☑ ● Variation 1 Specials (B)	🗁	325	15	4.62%	● 100%	69.5%

We used Google's Content Experiments technology, which is now built into Analytics. We determined a winning page (B) for this client after 455 visitor sessions with a 100 percent boost in leads at 70 percent probability. We used smart settings to increasingly show the probable winning page as it proved itself, which is why page B shows more sessions. With 70 percent likelihood that the B page would significantly outperform the original, we closed the experiment and deemed it successful.

Types of conversion goals you can test in Google Content Experiments

URL destination goals: An experiment that uses a URL destination goal focuses on getting users to view a specific web page. We tested for this goal in our case study, but the page was a thank-you page that visitors only saw if they completed a lead form, so what we were really testing was how many more would fill out the promotional lead form for the hotel.

Event goals: An experiment that uses an event goal focuses on getting users to perform a specific action on a page, such as signing up for a newsletter, viewing a video, or adding a product to an online shopping cart.

Session duration goals: This kind of goal tests to see how well a test page encourages users to spend a minimum amount of time on your site. For example, on a news site you can see how long visitors spend on articles and validate the rates charged for advertisements.

Pages-per-session goals: This sort of goal helps you understand whether users are consuming certain content. Are they reading product pages, looking at services, and reading your blog?

12

Review. Refine. Repeat.

The best business coaches include reminders in their teaching to "sharpen the saw" (Stephen Covey), and make continual improvements.
As you embrace testing in CRO, you can make the iterative improvements that will help define your career because they are essential to your business. You will also avoid "perfection paralysis".

Remember:
"Progressive improvement beats the hell out of postponed perfection." — Tim McEneny, Sr.

Chapter 12. Review. Refine. Repeat.

Chapter 1 - Takeaway

Getting people to visit your site is important, but it's crucial to keep them there long enough to convert them into leads. The best way to do this is through testing. Testing can be time consuming and even a bit tedious, but taking the time to create a site to which your target audience best responds will result in the maximum return on your investment.

Chapter 2 - Takeaway

In order for testing to reveal any meaningful results, you must first have a clear idea about your website's objectives. Once your objectives are defined, you can use CRO tests to help understand visitor behavior on your site.

Chapter 3 - Takeaway

Don't get so hung up on the details, semantics, and design elements of your site that you lose sight of your optimization goals. The important thing is for your site to be intuitively navigable and for your content to be both engaging and relevant.

Chapter 4 - Takeaway

It's important to pay attention to the intentions of your site visitors. Evaluating your interface options from the get-go can save you a lot of money in programming costs.

Chapter 5 - Takeaway

SEO is not the end-all, be-all. Rather than pepper your content with sensational keywords that will drive irrelevant traffic to your site, localize your vocabulary to secure a spot on the top of the search engine results pages that matter most to your business.

Chapter 6 - Takeaway

It's not just important to be clear about your site's objectives; the value of your conversion goals are also relevant.

Chapter 7 - Takeaway

Testing isn't just about version A versus version B. It's about creating a multitude of options that may involve only minor differences. Those differences may seem small at first glance, but they can be significant in influencing user behavior on your site.

Chapter 8 - Takeaway

When creating sign-up forms for your website, be sure to simplify and personalize the experience. Qualifying questions at the front of the form should determine what questions are relevant for your visitor in the remainder of the form.

Chapter 9 - Takeaway

Start your first A/B test for PPC ads, landing pages, and email subject lines. All three, ideally. Then refine and repeat as necessary.

Chapter 10 - Takeaway

You're not optimizing something inanimate like a landing page or a website, you're optimizing for a desired behavioral outcome. Paying attention to the psychology of motivation is a key to successful persuasion.

Chapter 11 - Takeaway

A/B testing can help increase conversions significantly, as illustrated by the A/B testing we did for *Croc's Casino Resort* in Costa Rica. Multivariate testing can be useful, but sometimes it's simpler and easier to distinguish what elements are making the most impact by electing to test only one or two elements. Headlines are the most important aspect to test, followed by CTAs.

Final Takeaway

How much does it cost to do CRO? We could probably write two or three chapters to answer that question, but since this is a concise guide, we'll leave you with this thought: in terms of lost revenue, it is less expensive in almost every case to do CRO than it is to *not* do CRO. This is because of the additional customers you win.

13

CRO Tips

"If your team is able to competently control conversions, your business can control the cost and effectiveness of all online expenditures, from advertising to traffic generation." — Brian Massey in "Your Customer Creation Equation"

---

13. 25 CRO Tips

1. Always be testing. Testing will directly and measurably affect your ROI, usually more quickly than other strategies and tactics.

2. Set goals. You can't measure what you haven't defined. In analytics, this means assigning dollar values to specific outcomes so you can identify the exact value of each action visitors perform on your site.

3. Measure results. Report results. Reporting results guides your team to make crucial adjustments to your campaigns and curries the favor of clients and bosses.

4. Remember that every campaign is different. Every landing page is different and every B2B or B2C prospect is different. Stay flexible and open to many options.

5. Cultivate outside-in thinking. Too often, entire sites and marketing campaigns are dictates of high-level executives and managers within an organization, rather than actual data based on user experiences and opinions.

6. All leads are NOT the same. Find the most valuable customers and convert them.

7. Test the button copy. "Click Here" and "Go" are the most successful terms for submit buttons, but how about getting a little more specific? Testing with more descriptive elements has the potential to yield better results.

8. Form length really does matter. If you want to increase form conversions, you must consider reducing the number of fields. Don't alienate potential customers with a ridiculously time-consuming form.

9. Don't call me. Forms that ask for phone numbers see their conversion rates dip an average of 5 percent.

10. Make your call to action prominent. Having your CTA above the fold on your site will eliminate loss of potential business for lack of clarity.

11. Analyze every page of your site. CRO is about process and flow. It's important to make sure that your entire site is both easily navigable and intuitive.

12. Offer promotions and discounts. Site visitors are heavily motivated to complete desired actions when there is a promotion involved.

13. See what the competition is up to. Looking into what your competitors are doing will give you a good sense of what works and what doesn't.

14. Test those email subject lines. Writing two complete emails with different graphics and CTAs does take quite a bit of time (although we believe it will pay off), but the least you can do is write two different subject lines to test which will compel more opens.

MailChimp, our email program of choice, makes this fast and easy.

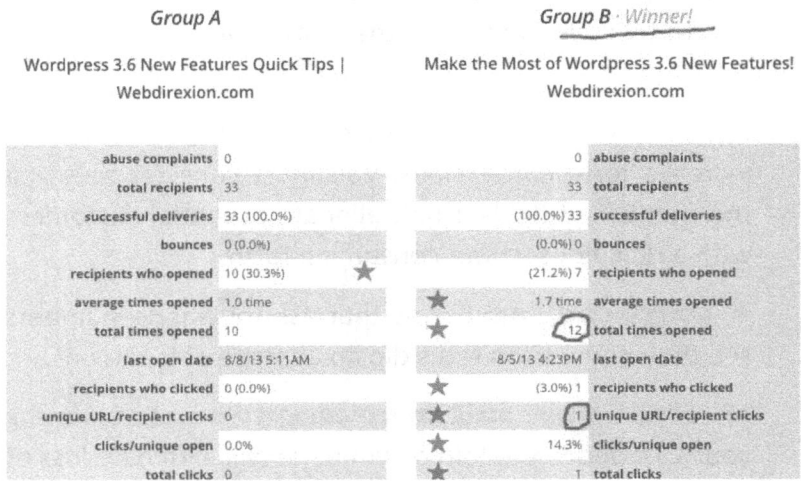

Group A	Group B · *Winner!*
Wordpress 3.6 New Features Quick Tips | Webdirexion.com	Make the Most of Wordpress 3.6 New Features! Webdirexion.com

Group A		Group B	
abuse complaints	0	0	abuse complaints
total recipients	33	33	total recipients
successful deliveries	33 (100.0%)	(100.0%) 33	successful deliveries
bounces	0 (0.0%)	(0.0%) 0	bounces
recipients who opened	10 (30.3%) ★	(21.2%) 7	recipients who opened
average times opened	1.0 time	★ 1.7 time	average times opened
total times opened	10	★ 12	total times opened
last open date	8/8/13 5:11AM	8/5/13 4:23PM	last open date
recipients who clicked	0 (0.0%)	★ (3.0%) 1	recipients who clicked
unique URL/recipient clicks	0	★ 1	unique URL/recipient clicks
clicks/unique open	0.0%	★ 14.3%	clicks/unique open
total clicks	0	★ 1	total clicks

Above is an example of an A/B test we did for an email subject line about a WordPress upgrade. Even though we sent it to a small group, MailChimp determined a winner, then sent the rest of the emails out with the winning subject line in place.

15. Offer two choices on opt-in pop-ups and forms (hat-tip, CopyHackers.com). In addition to asking for the action you want, offer an opt-out button too. Why? Persuasion psychology research shows that if you show a negative choice with consequences, your conversions will increase.

Try the forms choice tactic using two buttons:

YES - I want leads	NO leads for me

When using forms, try the negative consequence as a second-button tactic to increase conversions.

16. Use the Massey formula with elements for an effective landing page (see: ConversionScientist.com). "Effective landing page takes a Web Page (**Wp**), adds an Offer (**Of**), a Form (**Fm**), an Image (**I**) of the product plus Proof (**Pr**) and Trust (**Tr**)."

Brian Massey's work on conversions is first rate. Massey patterns this as a table of elements for conversion science and the end result looks like this:

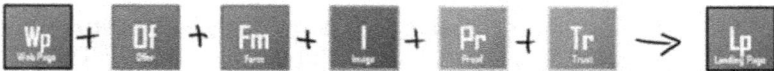

$$Wp + Of + Fm + I + Pr + Tr \rightarrow Lp$$

17. Understand what you are optimizing.

Remember that you're not optimizing a landing page, advertisement, email, or website for conversion rate increases. You're really optimizing for desired behavior. It's a subtle difference, but it's oh-so-important when it comes to the psychological motivation triggers that will push the buttons of your prospects.

18. Throw out the playbook. Don't listen to advice based on past tests or insider experience. No designer knows the best converting layout. No marketing director knows the best pitch and call to action every time. Every test is unique. The only input you want is from outside of your team and the client's team—that is, the insight you get from prospects and end customers.

19. Think in terms of iterative testing. This strategy means that you take insights from a previous test and use them in a sequence of new tests to ratchet up your success. Let's say that in your first A/B test, your B landing page offers a different headline, photo, and CTA. Once that page wins, you can quickly try another test where you change just the headline, or just the image, to look for an incremental lift in your results.

20. Forget iterative testing—sometimes (hat-tip, DigitalMarketer.com). Iterative testing usually assumes you will stay within the same overall design and just try a test on one of the page elements, such as the main headline. Sometimes the change gives little or no result. So, consider a radical and completely different page design for your next test. Apply the same Massey formula (see Tip 16) while also keeping in mind the psychology of your target personas. But take a radically different creative approach.

21. Solving pain points will motivate prospects to take action. There's that *persuasion psychology* thinking again. What are pain points? Well, one industry *Webdirexion* specializes in is travel and hospitality (more specifically marketing for hotels and inns). Most innkeepers and hoteliers have a limited amount of time (a pain point) because they wear lots of hats—don't we all? So we have to be careful when suggesting a new service to them and assure them that it will save time.

22. Use the Meclabs.com Formula: It's all about how people make choices. Dr. Flint McGlaughlin, a pioneer in CRO, has published a formula for success involving the following considerations: **M**otivation of user, **C**larity of the value proposition, **I**ncentive to take action, **F**riction elements of process, and **A**nxiety about entering information. The exact formula is:

$$C \text{ (probability of Conversion)} = 4M + 3V + 2(I - F) - 2A$$

23. Clarify your value proposition. Yep. It was in the last tip. But it's so important that it bears repeating. If prospects can't understand the value of your offer, your other CRO efforts will be fruitless. Common sense, right? You'd think so, but examples of hazy offers abound, even at companies of considerable size.

24. Test an appropriate number of people. When you start a test, you should have a firm grasp on the approximate number of test subjects to find a winner. Both Google Content Experiments and MailChimp do some sophisticated math to point toward a winner. Where do they get their formulas? Statistics. Just remember two things: the more variables you test, the more people you need in the test group, and PPC ads are usually a great way to both bring the number of new eyes you need while ensuring that you get qualified visitors to your test locations. A large mailing list helps, too. Formulas are out there for the Googling, but it's important to think in terms of several hundred to a few thousand for most scientifically valid tests.

25. Remember that CRO is *not* a precise science. Formulas can be incredibly useful, but remember that people buy on emotion, not bullet points (for more details, read Ogilvey—an iconic advertising figure on which the TV Show *Mad Men* was partially based). Advertising, focus group, and market research have proven this for many years. This means that you need to get creative in your landing page designs and use intuition in the CRO process. We've outlined some of the science; now use professional graphic design tactics in your next test.

26. Bonus Tip: Content Marketing + CRO = Connection. And when you have that emotional handshake with your prospect, they will make that final conversion to a Customer—the checkmate where you take the connection king, clear the board and start the next round in the testing game.

14

Resources: Tools & Books

You can't take credit for that which you don't test, other than to say you created something, but you're not sure how it is performing. Keep on learning, then testing and learning.

---

14. Resources: Tools and Books

Do you have a budget for testing? When you have one and use it wisely, it will more than pay for itself in terms of better leads and more of them, which will result in more sales.

Tools:

- **Google Analytics** - Google Analytics offers built-in A/B testing (sometimes called "split testing"). You can test up to 10 full versions of a single page, each delivered to users from a separate URL. Note that the more versions you use in a test, the more traffic you'll need to get a fix on the winning variation. This A/B testing method allows you to do your test on your site itself, after you create two page variations to test and insert some code for tracking results. There's also a WordPress plugin available. We use this a lot at *Webdirexion* and like the feature that selects the winning version of a page and runs it automatically.

- **Instapage.com** - This is a landing page builder that includes built-in A/B testing. It works great for external tests, and can be set to appear under your own domain name.

- **Unbounce.com** - A competitor to Instapage, this offers an intuitive page builder, A/B testing, and dynamic text replacement—a useful feature for when you are running different types of PPC ad campaigns.

- **CrazyEgg.com** - This includes heat maps, scroll maps, overlays, and confetti maps thatshow referral sources by color.

- **Marketizator.com** - This resource bills itself as a CRO tool, and includes four components: A/B testing; personalization; surveys; and segmentation. They also offer a *CRO tools comparison chart*

- **LuckyOrange.com** - Heat maps, recordings of visitor sessions, polls, and even a support chat engine (chat to your visitors about support, or about site visitor experience).

- **TrueReview.com** - Real clients meet your staff and then leave a review.

- **FiveSecondTest.com** - Quick but effective tests you can use creatively (we like this service for testing prior to coding).

> **Editor's Note:** We recommend you sign up to be a free tester with FiveSecondTest.com so you can learn what other marketers are optimizing and how they are using the system. But beware: testing can be addictive. When we got sucked in and ended up doing a dozen or so tests without even realizing that time was passing. By actively volunteering as testers, we found that the positioning of graphic elements made a huge impact on how we experienced pages and their intended messages.

- **NavFlow.com** - The same company behind Five Second Tests offers nav flow testing with volunteer testers from across the web.

- **ClickTale.com** - ClickTale provides not only heat maps, but also session feedback recordings and conversion funnel visualizations.

- **SurveyMonkey.com** - Testing will give you some great gains, but sometimes you need to ask visitors what's on their minds to discern intent.

- **Nelio.com** - A WordPress-based testing platform that uses a plugin that helps you test and review results for pages, posts, headlines, widgets, and themes using A/B tests, heat maps, and more

- **TreeJack** (Optimalworkshop.com/treejack) – This resource by Optimal Workshop analyzes your site navigation structure to determine user-friendliness.

- **Usabila.com** - Take a holistic approach to gaining insights from site visitors by combining several types of feedback with analytics integration, targeted questioning, and ClickTale integration. For the technically savvy, there is also integration with Webhooks.

- <u>VWO.com</u> - Short for Visual Website Optimizer, this tool offers standalone (in the cloud) or WordPress plugin versions for A/B testing. It also provides multivariate testing along with a number of additional tools, including behavioral targeting, heat maps, and usability testing.

Books:

Here are a few books that have helped us with our formula of three Cs: Content + Conversions = Connection.

- <u>Managing Content Marketing</u>: *The Real-World Guide for Creating Passionate Subscribers to Your Brand* (by Robert Rose and Joe Pulizzi). This book includes a thorough examination of target personas, and the journey storytellers should take to engage prospects. In it, co-authors Joe Pulizzi and Robert Rose really get the three-C formula. Even though their business and conferences focus on content, this book starts by acknowledging that, "Content marketing is useful beyond the customer stage—it can be used as an upsell or reinforcement mechanism." This occurs at the lead nurturing phase. In the back of the book, there is a chart that shows the Goal Pyramid, which includes A/B tests, PPC management, and a lot of the tactics we cover in this guide. To get the big picture, get this book.

- **Content Marketing Strategies for Professionals:** *How to Use Content Marketing and SEO to Communicate with Impact, Generate Sales and Get Found by Search Engines* (by Bruce Clay and Murray Newlands). The title tells you what to expect. Co-author Bruce Clay is internationally known for his SEO expertise and this book is somewhat of a bible for serious content marketing tactics. Content marketing is not all about SEO, and the authors know this: "The work of a marketer is to utilize the human needs for connection and story and to communicate messaging that compels people to not only listen, but also buy in and become customers."

- **Secret Formulas of the Wizard of Ads** (by Roy H. Williams). Arguably one of the best books on copywriting ever written, his book goes beyond copy content into the depths of the prospect's psyche and draws from both the sciences (there's a chapter on how auditory association meets visual association) and the arts ("The Ad Writer as Poet"). Just mastering a few of these skills will lead to more conversions.

- **Branding Basics for Small Business:** *How to Create an Irresistible Brand on Any Budget* (by Maria Ross). Branding comes before content, but some managers, business owners, and even C-level executives misunderstand what a brand actually is. Your landing pages, emails, and small text ads should serve your brand's messaging, and it's important to keep branding in mind during your work with the three Cs from landing page logos to slogans. Though she doesn't specifically address landing-page components, author Maria Ross makes clear what a brand is—and how to master branding.

- **Influence**: *The Psychology of Persuasion* (by Robert Cialdini). We mentioned this one earlier in the guide, and include it here because you need to understand your target persona's motivational psychology if you want to maximize conversions. This is not a book focused on content or the other Cs. Instead, Cialdini arms you with seven chapters of persuasion strategies that dive deep into motivational factors from single jump points such as reciprocation, liking, authority, scarcity, and social proof. You need to consider this thinking at the heart of your creative conceptions so that you base your conversion pages and elements on what moves people.

- **Advanced Google Adwords** (by Brad Geddes). We move more into the science of conversion with this book, but the content of ads is still a core component. The book does a good job of focusing on optimizing for conversions, with a look at tactics for doing so for both ads and landing pages (which must be in sync to get a good quality score by Google).

- **Email Persuasion**: *Captivate and Engage Your Audience, Build Authority and Generate More Sales With Email Marketing* (by Ian Brodie). We talked about A/B testing emails earlier, but how exactly do you get more emails read by your prospects? Brodie delivers that answer among other gems in this book. He points out that email provides a way to have a personal touch with thousands of prospects and clients. Study this quote carefully (then get the book): "The ability to communicate proactively, personally, and regularly makes email marketing an incredibly powerful tool for building relationships…"

- **Landing Page Optimization** - *The Definitive Guide to Testing and Tuning for Conversions* (by Tim Ash). Mr. Ash says there are three keys to online marketing: Acquisition, Conversion, and Retention. Those parallel nicely our three Cs, with conversion being central to the book's focus. It takes an intense look at types of landing pages, page elements and "tuning methods." Ash, like all great marketers, knows that you begin with understanding your audience and he devotes an entire chapter to just that.

- **Your Customer Creation Equation** - *Unexpected Website Formulas of The Conversion Scientist™* (by Brian Massey). Are you ready to dive deeper into the science of conversions? If so, this is your book. Massey makes many great tactical and strategic recommendations writing about how most people start build a landing page within their existing site's template layout. We particularly liked this gem: "... your page already has marginally relevant distractions built in. Forget all that. Start with a blank page."

- **eMarketing Strategies for the Complex Sale** (by Ardith Albee). Too many web teams miss the vital connection between internal teams and external customers, and with this understanding, Albee brings a fine detail to processes that contribute to online marketing success. These range from strategies for dealing with buyer personas to lead nurturing to meaningful metrics. You'll find some valuable hands-off tactics used to communicate key sales metrics between the traditionally siloed marketing and sales teams. Details on marketing impact scoring in the context of sales are also provided.

- **Mastering the Complex Sale:** *How to Compete and Win When the Stakes are High* (by Jeff Thull). In the introduction to this guide, we made the point that a lot of marketers stay focused on the first C in the process: content. They shouldn't be so single-minded, and this book illustrates why. We maintain that if you do not understand the connection process—traditionally fulfilled by a sales team in B2B selling—then you will not be as persuasive as you could be. You will also not be able to deliver the best qualified leads from your landing pages. The best sales people need to communicate complex solutions in B2B marketing. The best CRO teams will understand that and work in sync with sales teams to provide better qualified leads and insights that salespeople can use to close more details.

Appendices

Exercises, Plugin Reviews, and More

He who learns but does not think, is lost! He who thinks but does not learn is in great danger.
— Confucius

Learning never exhausts the mind.
— Leonardo da Vinci

I like to listen. I have learned a great deal from listening carefully. Most people never listen.
— Ernest Hemingway

Appendix I: CRO Exercises

Exercise 1 – Brainstorm a landing page test:
Describe persona(s) you wish to persuade (*Chapt. 10*):

_____

_____

Buyer's journey: ❏ Awareness ❏ Comparing ❏ Deciding
(see: Chapt. 10 – gear your page to the buyer's journey phase)

HEADLINE(s):	Quickly write a couple of headlines that solve pain points.
A: _____ B: _____	
What image(s) persuade emotionally? A: _____ B: _____	Review pro-stock images—find two to test.
COPY: Value proposition notes:	Make your value proposition clear.
Social Proof:	This could be a form—what are two approaches you'd like to test?
Calls to Action: A: _____ B: _____	
Overcome Anxiety:	What can you include to relieve a prospects' nervousness about giving you their email info.?
Concluding Brand Info/Logo	

Exercise 2 - Brainstorm a **Call to Action Form** Test:

List the persona(s) you wish to persuade:

_____

_____

State call to action: a) _____ b) _____	Write a couple of compelling form headlines.
First "qualifier" question? _____ ❏ Drop-Down? ❏ Check Boxes? **Second "qualifier" question?** _____ ❏ Drop-Down? ❏ Check Boxes?	Write a couple of qualification questions to make sure. Gather their contact information without too much "friction."
❏ Name | ❏ eMail ❏ Phone | ❏ Company ❏ Other: _____	In this section think about how to make your form "smart" (see chapter 8) with conditional logic. For example, if one of your qualifiers above is budget, you may show different solutions from which to select.
Conditional logic ideas: _____ _____	
Include a comments field?	
Concluding brand info./logo	

Exercise 3 - Calculate Goal Value

Name your conversion goal (ie. download a PDF, fill out lead form): _____

Ave. Sale Amount: $ _____

Estimate the average value of your product or service.

+

Secondary value: $ _____

Repeat purchase value.

=

Total acquisition value: $ _____

Add these two values.

÷

Leads per sale: _____

Divide by Leads per sale (How many leads does it take before your sales team closes a sale?)

=

Goal value: $ _____

Enter the last number into Google Analytics as your $ value for this goal.

EXAMPLE: You have a service you sell for $1000, and 80 percent of customers buy another service costing $500 ($400 is 80 percent of $500). Your total acquisition value is $1400 to your company. Your sales team can close one in 10 of the leads you deliver at your current qualification levels. You divide $1400 by 10 and get a goal value of $140 to enter into Google analytics.

Appendix II: 4 CRO Service Reviews

While researching this guide, we came across G2Crowd.com and found they had crowd-researched poplular A/B and conversion testing services, as detailed in the following chart (used with permission):

Optimizely	**VWO**	**Google Analytics**	**unbounce**
★★★★☆	Visual Website Optimizer	★★★★★	★★★★☆
(40 ratings)	★★★★★	(166 ratings)	(21 ratings)
	(34 ratings)		
Optimizely.com	**VWO.com**	**Google Analytics**	**UnBounce.com**

	Optimizely.com	VWO.com	Google Analytics	UnBounce.com
Standard Targeting - Optimization				
	8.2	8.9	8.7	7.6
	Based on 20 answers	Based on 33 answers	Based on 24 answers	Based on 19 answers
Custom Targeting - Optimization				
	8.2	8.4	8.5	8.2
	Based on 20 answers	Based on 33 answers	Based on 24 answers	Based on 19 answers
Profile Storage Duration - Optimization				
	8.5	9.3	9.0	6.9
	Based on 20 answers	Based on 33 answers	Based on 24 answers	Based on 18 answers
A/B Testing - Optimization				
	9.6	9.0	8.2	9.5
	Based on 20 answers	Based on 33 answers	Based on 24 answers	Based on 19 answers
Conversion Goal - Optimization				
	9.1	8.7	8.6	8.8
	Based on 20 answers	Based on 33 answers	Based on 24 answers	Based on 19 answers
Percentage of Traffic - Optimization				
	9.3	8.9	8.7	9.2
	Based on 20 answers	Based on 33 answers	Based on 24 answers	Based on 19 answers
Confidence Level - Optimization				
	8.5	8.9	7.5	9.0
	Based on 20 answers	Based on 33 answers	Based on 24 answers	Based on 19 answers
Multivariant Testing - Optimization				
	8.4	8.7	7.9	9.3
	Based on 20 answers	Based on 33 answers	Based on 24 answers	Based on 19 answers
Split URL testing - Optimization				
	9.0	8.8	7.8	7.8
	Based on 20 answers	Based on 32 answers	Based on 24 answers	Based on 19 answers

Copyright G2 Crowd, Inc. More at: G2Crowd.com (see "all categories").

You'll find still more review criteria for these four services along with many other software reviews at the G2Crowd site.

Comparison of Starter Plans for Same Four Services

Optimizely:	VWO:	Google:	Unbounce:
$49 mo.	$49 mo.	1 plan - Free	$49 mo.
One "Project"*	Unlimited "Campaigns"	Unlimited "Experiments"	Unlimited "Pages"
Unlimited Users	Five Users	Unlimited Users	Single User
heat maps	heat maps	No heat maps	No heat maps
Page Builder	Page Builder	No Page Builder	Page Builder
3rd Party Integrations	Third-Party Integrations	N/A	3rd Party Integrations

Optimizely "project" includes "one website, one iOS App, and one Android App". Above table compiled by the CRO Guide editors.

Our concise take? Total G2 Review scores are close, and you can do a good test with any of these services. All start at $50/month, except for Google Content Experiments (a part of Google Analytics), which is free. All except Google Analytics also have landing page-creation tools. Test scoring methodologies will vary depending on each system's algorithms. We recommend you test two-three services to determine which is the right fit for your team; each site has posted a longer list of features for each plan.

Glossary of Acronyms

B2B – Business to Business

B2C – Business to Consumer

CMI – Content Marketing Institute

COMP – College of Marketing Professionals

CRM – Customer Relationship Management

CRO – Content Rate Optimization

CTA – Call to Action

KPI – Key Performance Indicator

LPO – Landing Page Optimization

PPC – Pay Per Click

ROI Return on Investment

SEO – Search Engine Optimization

SERPs – Search Engine Results Pages

TLA – Three-letter Acronyms

VBO – Visitor Behavior Optimization

CRO RESOURCE Guide

In this section we present services, courses, & products for conversion rate optimizers

- Nelio A/B plus Heatmap testing for WordPress Websites - free trial
- Webdirexion 1-2-3 Forms for smart CRO - 15% Off
- College of Marketing Pros (COMP) Marketing Classes with Special Discounts
- Promoted Stories from Dlvr.it - try for $1
- WP Conversion Boxes Pop-ups Plugin - Save on Five
- CRO Power Online Course - Sign-up Discount
- WordPress Content Marketing Power Course - Sign-up Discount
- Checkfront... a powerful way for Inns & Hotels to capture bookings and optimize results - Free Trial
- OnePageCRM + Unbounce for a powerful 1-2 CRO punch - Free Trial
- MailChimp Content Content A/B Testing - Free for up to 20,000 emails a month

These listings are paid for or made available via direct marketing agreements with the publisher. See last page for more details.

Nelio A/B Testing Platform for WordPress

CRO Tip: Most marketers think just about testing landing pages. Instead, test pages, *plus* posts, products, headlines, widgets, CSS variations, themes, and menus.

Test your whole WordPress site.

Heatmaps

Run Multiple Experiments

Results at a Glance!

Detailed Results

NELIO A/B TESTING *for WordPress*

Nelio A/B Testing is the most powerful and versatile conversion optimization service for WordPress. It helps you define, manage, and keep track of A/B-testing experiments, combined with powerful and beautiful Heatmaps. Nelio's also compatible with WooCommerce.

In particular, you can test alternative names, featured images, and descriptions for your products, and use your orders as conversion actions (so that you can make sure that a certain product has been purchased).

Now… enjoy an "unlimited*" Free Trial
Webdirexion.com/Nelio-A-B-Testing

*Up to 1,000 page views - no time limitation. Includes support.

Webdirexion 1-2-3 Smart CRO Forms

CRO Tip: Try the triple-play with conditional logic in our forms. First, show or hide fields based on how your prospects answer questions; send a custom reply based on their answers; and finally, redirect them to different custom thank you pages with customized content.

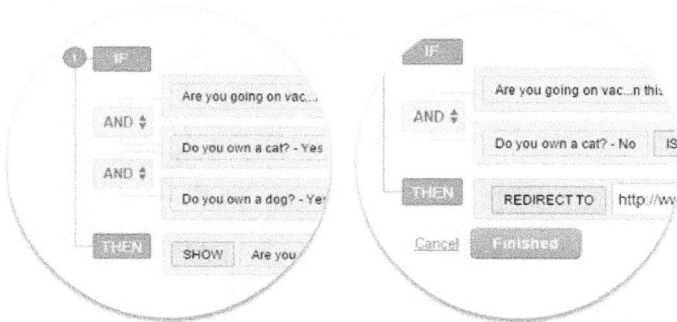

Webdirexion **1-2-3 Forms** include technical support and a robust set of integrations and features, including capabilities to accept payments on your site, publish forms anywhere (including all major CMS platforms and Facebook), develop multi-page forms, translate forms into multiple languages, create polls and quizzes, and more.

15% Off: Forms are included in all *Webdirexion* WordPress Coverage Plans, or order them stand-alone for $19.95/mo. (one annual payment of $239). Take **15% OFF** by using promo code: **SAVE-15-CRO** on your order form.

Webdirexion.com/WordPress-Support-Service-Plans

COMP - Now Practicing CRO

CRO Tip: At COMP—the College Of Marketing Pros—we practice CRO in a number of ways, including using a directory plugin that lets us move popular courses to featured positions at the top of the list:

COMP is an online directory of top marketing courses for busy working professionals. Find courses for marketing managers at Udemy, an e-learning platform that provides lifetime access via your computer or mobile apps.

Save: Go to the COMP Specials page — CollegeofMarketingPros.org/specials - and enjoy **30% off** selected courses when you select "CRO-Guide" under "How did you hear about us?" Our smart 1-2-3 form from Webdirexion will mail the exclusive special only to those who select CRO-Guide.

Dlvr.it Promoted Stories for Social CRO

CRO Tip: Run a promoted story via the Dlvr.it platform, then track the results from each social media network using your google analytics program:

Dlvr.it delivers tracking stats for your promoted stories then use Google Analytics to track conversions by social network.

Boost website traffic and leads by distributing your blog posts to search, social, mobile and local media. Get your content in front of interested readers via thousands of social news feeds and news sites. Reach millions of readers of The New York Times, cNet, Washington Post, ABC-local and more.

Special - Promote your first story for $1: Give it a try today for $1. Webdirexion.com/PromotedStories

CRO Power Online Course for CRO Guide Readers...

CRO Tip: Incentives. Figure this one concept out and you will have unleashed a powerful motivator to compel more conversions—we have one for you below.

Go beyond A/B testing to learn advanced tips, tactics and techniques in CRO for the working marketing professional.

The CRO Power online course is the official companion course for this book. In the course, CRO Guide author Scott Frangos reviews the key conversion improvement concepts and exercises in this book. Take & refer to this course any time, on any device—with lifetime access!

Save $50! - The publisher and author have arranged a $50 savings on this companion course for readers of the CRO Guide and students of marketing. Just use the URL below to sample the course and receive your discount!

<u>Webdirexion.com/cro-power-course</u>

Checkfront — the ultimate booking software for savvy marketers

CRO Tip: Get conversions (reservations) through other websites! Use Checkfront's embeddable widget to place your booking engine on allied business websites.

Checkfront
Smart, Simplified Online Bookings

A new way to manage your business that takes the hassle out of reservations & online payments.

View Bookings	Adjust Inventory	Export & Update	Customer Insight
Get a birds eye view of your bookings and customers. Make better business decisions based on easy-to-understand statistics.	Dynamically adjust your room inventory and availability. Set up seasonal package deals and sale prices, as well as group discounts.	Quickly update and export bookings to Excel, CSV, or PDF formats for easy viewing. Review & look for conversion rate optimization.	Know where your customers are coming from. Use analytics to pinpoint your audience and keep them coming back for more.

With over 50 integrations, including MailChimp, Facebook & TripAdvisor, there's no end to the ways Checkfront helps you optimize.

Now, managers of hotels & inns, tour companies, and rental properties have a simple solution to complex problems. Checkfront software makes it easy to manage bookings and customers through a unified tool set, while seamlessly integrating into popular services that empower your business. We have add-ons for CRMs, QR codes, Accounting & more.

Click below for a **Free 21-Day Trial**

Webdirexion.com/Checkfront

WordPress Content Marketing Power Course

CRO Tip: Studying what visitors do and don't do at your site—via goals you set to measure outcomes—helps you develop the best approach to using WordPress tactics.

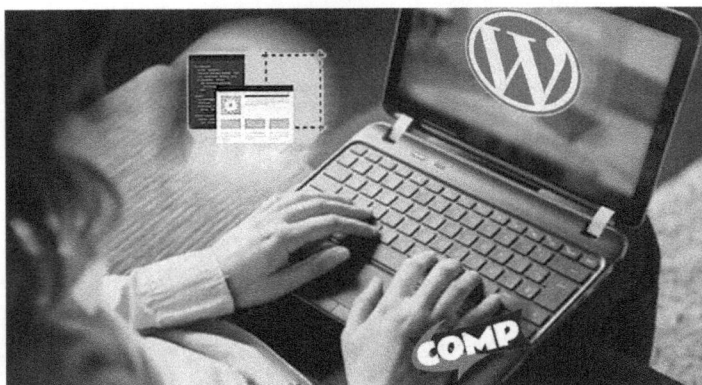

Join content marketer and WordPress developer Scott Frangos for a journey through plugins and advanced tactics.

Enroll once and enjoy lifetime access to this course and its regular updates, with new lectures and over six hours of video. Watch lectures any our of the day on any web-enabled device. This course helps students understand the principles and tactics of marketing strategy and how to use WordPress to best enable those tactics at your sites. Topics include SEO and a review of six content marketing plugins for WordPress websites.

Save $40 - Special for CRO-Guide reader: save $40 when you you register using the URL below!

Webdirexion.com/WP-Content-Marketing

WP Conversion Boxes - A/B Testing

CRO Tip: In addition to A/B testing your Conversion Box popups, try a "scroll triggered" pop-up keyed to appear when a visitor reads a related content item on a page.

Create & Test Multiple Popups

Unique Visitors	Pageviews	Box Views	Conversions	Conversion Rate
56	69	44	3 Clicks	6.81%
55	72	46	6 Clicks	13.04%
1212	2105	1502	25 Optins	1.66%
980	1566	543	31 Optins	5.71%

Integrates with:

salesforce · pardot · AWeber · GetResponse
Constant Contact · Campaign Monitor
iContact · MailChimp
mailpoet · Infusionsoft

WordPress Plugin for CRO

Conversion Box plugin integrates with the best email providers — also insert html code for WuFoo and 1-2-3 Forms.

WP Conversion Boxes Pro is the most powerful WordPress plugin for creating, conversion tracking and A/B testing beautiful Email Opt-in and Call-to-Action Boxes on your WordPress Blog. Includes Timed Popups; Scroll-triggered Popups; Exit Popups; plus 60+ conversion optimized templates. Place your boxes everywhere - under individual posts, pages and custom post types, in the sidebar, on the top of your homepage (feature box), inside the content using a shortcode!

Special - $27 for one site — save on the Professional plan... 5 sites for $37 -
Webdirexion.com/WP-Conversion-Boxes

NEW MailChimp Content A/B Testing

CRO Tip: Test both subject lines (for more opens) and email content (for click-throughs). For larger lists, try tests for up to three versions of your campaign.

Will the use of different templates, content, or CTAs affect subscriber engagement? What subject lines are most effective? Does the time or day an email is sent affect the click rate? Test all of these variables.

Test different text blocks, images, links, CTAs, design elements, or even completely different templates to identify the most effective combination for your audience. MailChimp will determine the winner and then send the rest of your campaign to the winning combination. *Mailchimp is free for up to 12,000 emails to 2000 subscribers per month*:

Webdirexion.com/MailChimpTesting

(Mention the CRO Guide for **10% off** on a MailChimp Drip Marketing creative package from Webdirexion)

The 1-2 CRO Punch From OnePageCRM

CRO Tip: Immediately unite your web marketing with your sales team's efforts by tying lead forms to CRM systems. Leads are gathered into your CRM so the sales team can qualify each lead then nurture them and close them.

By integrating with Unbounce, incoming leads are shown on the top of your Action Stream for fast processing and increased conversion rates.

By applying productivity principles to CRM, we have turned complexity into a simple and focused to-do list. Finally, your team can focus on sales, not software. Plus we're integrated with MailChimp, Unbounce, Skype, Gmail, Xero Accounting, Evernote and more for a powerful, time saving working set up. Get more done.

Use the link below for a **Free Trial**

Webdirexion.com/OnePageCRM

*The CRO Guide listings are open to any company with a CRO related service or product. We ask that you include a CRO tip to help our readers and marketing students, and that there be a related special offer. For inquiries about a listing in the next edition of the Guide, contact the publisher, Webdirexion Publishing, at **888.974.9522**.*

Disclosure note about our CRO Guide Listings in the preceding section: Webdirexion Publishing, as a division of Webdirexion LLC, the online marketing agency, has affiliated marketing relationships with some of the companies with listings in this guide, and also mentioned elsewhere in the book chapters. Some listings may include direct offers from for services offered by Webdirexion, or from the author as is the case with courses taught by Scott Frangos. Not all offers herein are exclusive to this Guide.

Neither the author, publisher, nor Webdirexion LLC makes any warranty, express or implied, for performance or results of the services listed in this Guide. In the end your own results will depend on a wide set of variables including the quality of what you are offering, your creative strategies and tactics, your pricing considerations, and your position in the marketplace.

All service performance is the responsibility of each independent company and your agreements are directly with each company. Support is available directly from each company. In other words, while we've used a number of these services, and we like them, we encourage you to review alternative services and perform due diligence in selecting services and products to help in your conversion testing marketing efforts.

About the Author

Scott Frangos • Scott@Webdirexion.com • 888.974.9522, ext.700

Scott is a career marketing communications professional with over 20 years of college-level instruction experience. He is the founder and president of *Webdirexion LLC*, an online marketing agency with a heavy focus on CRO, writes articles about online marketing for the *Webdirexion* blog and has spoken at a number of Content Marketing conferences.

Editors & Contributors

Margot Hall, Managing Editor

Margot@Webdirexion.com • 888.974.9522, ext.703

Margot is a writer and content specialist who thrives on turning ideas into tangible results. She has worked with companies around the world, helping them develop strategies to both analyze and augment the impact of their digital content.

Whitney Beyer, Copy Editor

Whitney@Webdirexion.com

Whitney is a wordsmith and storyteller with a knack for brevity and authenticity. With extensive experience in student journalism and a current gig in PR and social marketing, Whitney employs her diverse skill set and can-do attitude to help others communicate their messages clearly and effectively.

Miranda Booher, Contributor

Miranda@Webdirexion.com • 888.974.9522, ext.708

Miranda specializes in the creation of compelling and engaging copy to drive customer acquisition and sales conversions. She's also an experienced entrepreneur who managed a landscaping business for several years before venturing into a career as a copywriter and content marketer.

Sherri Gutierrez, Contributor

Sherri@Webdirexion.com • 888.974.9522, ext.705

Sherri is a results-driven marketing professional with a lifetime focus on client satisfaction. She listens attentively to client needs and delivers strong solutions that increase profitability.

Julie Hume, Contributor

Julie@Webdirexion.com

Julie is a career marketing and communications professional, specializing in writing, SEO, and social media strategy. She was formerly the director of communications for *Amari Hotels and Resorts* and the content director for *eThailand*.

Special thanks to Michael Procopio, author of *42 Rules for B2B Social Media Marketing*, who read our CRO Guide with a critical eye and gave recommendations that made it better.

The Marketer's Concise Guide to CRO is a quick read covering tools, tactics and techniques to use conversion rate optimization to gather more qualified leads.

"...as I read, I took notes for a half dozen tests my clients *deserve*. Thanks, Scott, for the motivation and tools to be better at my job. — Joe Hage, CEO, *Medical Marcom*

"My biggest takeaway is the need to test every pitch, every call to action and every layout. Not even the most talented, experienced creatives know what will work and what won't. This book will make you rich." — Bob Leonard, Managing Consultant, *acSellerant*

"I'd recommend the CRO guide to anyone who wants to learn how to boost 'connections' with visitors. This guide will help you create more engaged readers and turn them into sales. In the end you will become a smarter marketer." — Bill Flitter, CEO, *Dlvr.it*

"I now feel way more confident that we can improve our CRO efforts. The book is a solid read and a quick one as well. Highly recommended! — Jon Wuebben, CEO *Content Launch*

Publisher: Webdirexion Publishing, a division of Webdirexion LLC
ISBN-13: 978-0-9969504-0-4 | **ISBN-10**: 0996950400
Price. $17.99 paperback | $7.99 Kindle and eBook

Webdirexion PUBLISHING

ISBN 978-0-9969504-0-4

9 780996 950404

51799>

www.ingramcontent.com/pod-product-compliance
Lightning Source LLC
Chambersburg PA
CBHW022057210326
41519CB00054B/614